Marx Joyce
Defoe Abbott Hardy Machiavelli Cooper Emerson Austen
Melville Montaigne Chesterton Hugo Grimm
Stoker Carroll Christie Haggard Eliot
Wilde Maupassant Byron Molière
Garnett Fitzgerald Engels Schiller
Goethe Einstein Hawthorne Smith Kafka
Cotton Dostoyevsky
Baum Henry Kipling Doyle Hall Willis
Leslie Dumas Flaubert Turgenev Nietzsche Balzac Crane
Stockton Vatsyayana Verne
Burroughs Whitman Gogol Vinci
Curtis Tocqueville Busch
Homer Widger Tolstoy
Darwin Thoreau Twain
Potter Freud Zola Scott Harte
Kant Jowett Stevenson Lawrence Dickens Plato
Andersen Burton Hesse
London Descartes
Poe Aristotle Wells Cervantes Voltaire Cooke
Hale James Hastings
Bunner Shakespeare Irving
Richter Chambers
Doré Swift Dante Chekhov Shaw Ida Benedict Alcott
Pushkin Wodehouse
Newton

⊕ tredition®

tredition was established in 2006 by Sandra Latusseck and Soenke Schulz. Based in Hamburg, Germany, tredition offers publishing solutions to authors and publishing houses, combined with world-wide distribution of printed and digital book content. tredition is uniquely positioned to enable authors and publishing houses to create books on their own terms and without conventional manufacturing risks.

For more information please visit: www.tredition.com

TREDITION CLASSICS

This book is part of the TREDITION CLASSICS series. The creators of this series are united by passion for literature and driven by the intention of making all public domain books available in printed format again - worldwide. Most TREDITION CLASSICS titles have been out of print and off the bookstore shelves for decades. At tredition we believe that a great book never goes out of style and that its value is eternal. Several mostly non-profit literature projects provide content to tredition. To support their good work, tredition donates a portion of the proceeds from each sold copy. As a reader of a TREDITION CLASSICS book, you support our mission to save many of the amazing works of world literature from oblivion. See all available books at www.tredition.com.

Project Gutenberg

The content for this book has been graciously provided by Project Gutenberg. Project Gutenberg is a non-profit organization founded by Michael Hart in 1971 at the University of Illinois. The mission of Project Gutenberg is simple: To encourage the creation and distribution of eBooks. Project Gutenberg is the first and largest collection of public domain eBooks.

Studies of American Fungi. Mushrooms, Edible, Poisonous, etc.

George Francis Atkinson

Imprint

This book is part of TREDITION CLASSICS

Author: George Francis Atkinson
Cover design: Buchgut, Berlin – Germany

Publisher: tradition GmbH, Hamburg - Germany
ISBN: 978-3-8472-2619-2

www.tredition.com
www.tredition.de

STUDIES OF AMERICAN FUNGI

MUSHROOMS

EDIBLE, POISONOUS, ETC.

BY

GEORGE FRANCIS ATKINSON

Professor of Botany in Cornell University, and Botanist of the
Cornell University Agricultural Experiment Station

Recipes for Cooking Mushrooms, by Mrs. Sarah Tyson Rorer
Chemistry and Toxicology of Mushrooms, by J. F. Clark

WITH 230 ILLUSTRATIONS FROM PHOTOGRAPHS BY THE
AUTHOR, AND COLORED PLATES BY F. R. RATHBUN

SECOND EDITION

NEW YORK

HENRY HOLT AND COMPANY

1903

[Pg ii]

INTRODUCTION. [Pg iii]

Since the issue of my "Studies and Illustrations of Mushrooms," as Bulletins 138 and 168 of the Cornell University Agricultural Experiment Station, there have been so many inquiries for them and for literature dealing with a larger number of species, it seemed desirable to publish in book form a selection from the number of illustrations of these plants which I have accumulated during the past six or seven years. The selection has been made of those species representing the more important genera, and also for the purpose of illustrating, as far as possible, all the genera of agarics found in the United States. This has been accomplished except in a few cases of the more unimportant ones. There have been added, also, illustrative genera and species of all the other orders of the higher fungi, in which are included many of the edible forms.

The photographs have been made with great care after considerable experience in determining the best means for reproducing individual, specific, and generic characters, so important and difficult to preserve in these plants, and so impossible in many cases to accurately portray by former methods of illustration.

One is often asked the question: "How do you tell the mushrooms from the toadstools?" This implies that mushrooms are edible and that toadstools are poisonous, and this belief is very widespread in the public mind. The fact is that many of the toadstools are edible, the common belief that all of them are poisonous being due to unfamiliarity with the plants or their characteristics.

Some apply the term mushroom to a single species, the one in cultivation, and which grows also in fields (*Agaricus campestris*), and call all others toadstools. It is becoming customary with some students to apply the term mushroom to the entire group of higher fungi to which the mushroom belongs (*Basidiomycetes*), and toadstool is regarded as a synonymous term, since there is, strictly speaking, no distinction between a mushroom and a toadstool. There are, then, edible and poisonous mushrooms, or edible and poisonous toadstools, as one chooses to employ the word.

A more pertinent question to ask is how to distinguish the edible from the poisonous mushrooms. There is no single test or criterion,

8

like the "silver spoon" test, or the criterion of a scaly cap, or the presence of a "poison cup" or "death cup," which will serve [Pg iv] in all cases to distinguish the edible from the poisonous. Two plants may possess identical characters in this respect, i. e., each may have the "death cup," and one is edible while the other is poisonous, as in *Amanita cæsarea*, edible, and *A. phalloides*, poisonous. There are additional characters, however, in these two plants which show that the two differ, and we recognize them as two different species.

To know several different kinds of edible mushrooms, which occur in greater or less quantity through the different seasons, would enable those interested in these plants to provide a palatable food at the expense only of the time required to collect them. To know several of the poisonous ones also is important, in order certainly to avoid them.

The purpose of this book is to present the important characters which it is necessary to observe, in an interesting and intelligible way, to present life-size photographic reproductions accompanied with plain and accurate descriptions. By careful observation of the plant, and comparison with the illustrations and text, one will be able to add many species to the list of edible ones, where now perhaps is collected "only the one which is pink underneath." The chapters 17 to 21 should also be carefully read.

The number of people in America who interest themselves in the collection of mushrooms for the table is small compared to those in some European countries. The number, however, is increasing, and if a little more attention were given to the observation of these plants and the discrimination of the more common kinds, many persons could add greatly to the variety of their foods and relishes with comparatively no cost. The quest for these plants in the fields and woods would also afford a most delightful and needed recreation to many, and there is no subject in nature more fascinating to engage one's interest and powers of observation.

There are also many important problems for the student in this group of plants. Many of our species and the names of the plants are still in great confusion, owing to the very careless way in which these plants have usually been preserved, and the meagerness of recorded observations on the characters of the fresh plants, or of the different stages of development. The study has also an important relation to agriculture and forestry, for there are numerous species which cause decay of valuable timber, or by causing "heart rot" entail immense losses through the annual decretion occurring in standing timber.

If this book contributes to the general interest in these plants as [Pg v] objects of nature worthy of observation, if it succeeds in aiding those who are seeking information of the edible kinds, and stimulates some students to undertake the advancement of our knowledge of this group, it will serve the purpose the author had in mind in its preparation.

I wish here to express my sincere thanks to Mrs. Sarah Tyson Rorer for her kindness in writing a chapter on recipes for cooking mushrooms, especially for this book; to Professor I. P. Roberts, Director of the Cornell University Agricultural Experiment Station, for permission to use certain of the illustrations (Figs. 1–7, 12–14, 31–43) from Bulletins 138 and 168, Studies and Illustrations of Mushrooms; to Mr. F. R. Rathbun, for the charts from which the colored plates were made; to Mr. J. F. Clark and Mr. H. Hasselbring, for the Chapters on Chemistry and Toxicology of Mushrooms, and Characters of Mushrooms, to which their names are appended, and also to Dr. Chas. Peck, of Albany, N. Y., and Dr. G. Bresadola, of Austria-Hungary, to whom some of the specimens have been submitted.

Geo. F. Atkinson,

Cornell University.

Ithaca, N. Y., October, 1900.

SECOND EDITION.

In this edition have been added 10 plates of mushrooms of which I did not have photographs when the first edition was printed. It was possible to accomplish this without changing the paging of any of the descriptive part, so that references to all of the plants in either edition will be the same.

There are also added a chapter on the "Uses of Mushrooms," and an extended chapter on the "Cultivation of Mushrooms." This subject I have been giving some attention to for several years, and in view of the call for information since the appearance of the first edition, it seemed well to add this chapter, illustrated by several flashlight photographs.

G. F. A.

September, 1901.

CHAPTER I.

FORM AND CHARACTERS OF THE MUSHROOM.

Value of Form and Characters. — The different kinds of mushrooms vary in form. Some are quite strikingly different from others, so that no one would have difficulty in recognizing the difference in shape. For example, an umbrella-shaped mushroom like the one shown in Fig. 1 or 81 is easily distinguished from a shelving one like that in Fig. 9 or 188. But in many cases different species vary only slightly in form, so that it becomes a more or less difficult matter to distinguish them.

In those plants (for the mushroom is a plant) where the different kinds are nearly alike in form, there are other characters than mere general form which enable one to tell them apart. These, it is true, require close observation on our part, as well as some experience in judging of the value of such characters; the same habit of observation and discrimination we apply to everyday affairs and to all departments of knowledge. But so few people give their attention to the discrimination of these plants that few know the value of their characters, or can even recognize them.

It is by a study of these especial characters of form peculiar to the mushrooms that one acquires the power of discrimination among the different kinds. For this reason one should become familiar with the parts of the mushroom, as well as those characters and markings peculiar to them which have been found to stamp them specifically.

Parts of the Mushroom. — To serve as a means of comparison, the common pasture mushroom, or cultivated form (*Agaricus campestris*), is first described. Figure 1 illustrates well the principal parts of the plant; the cap, the radiating plates or gills on the under side, the stem, and the collar or ring around its upper end.

The Cap. — The cap (technically the *pileus*) is the expanded part of the mushroom. It is quite thick, and fleshy in consistency, more or less rounded or convex on the upper side, and usually white in color. It is from 1–2 cm. thick at the center and 5–10 cm. in diameter. The surface is generally smooth, but sometimes it is torn up more or less into triangular scales. When these scales are prominent they are often of a dark color. This gives quite a different aspect to the plant, and has led to the enumeration of several [Pg 2] varieties, or may be species, among forms accredited by some to the one species.

The Gills.—On the under side of the pileus are radiating plates, the gills, or *lamellæ* (sing. *lamella*). These in shape resemble somewhat a knife blade. They are very thin and delicate. When young they are pink in color, but in age change to a dark purple brown, or nearly black color, due to the immense number of spores that are borne on their surfaces. The gills do not quite reach the stem, but are rounded at this end and so curve up to the cap. The triangular spaces between the longer ones are occupied by successively shorter gills, so that the combined surface of all the gills is very great.

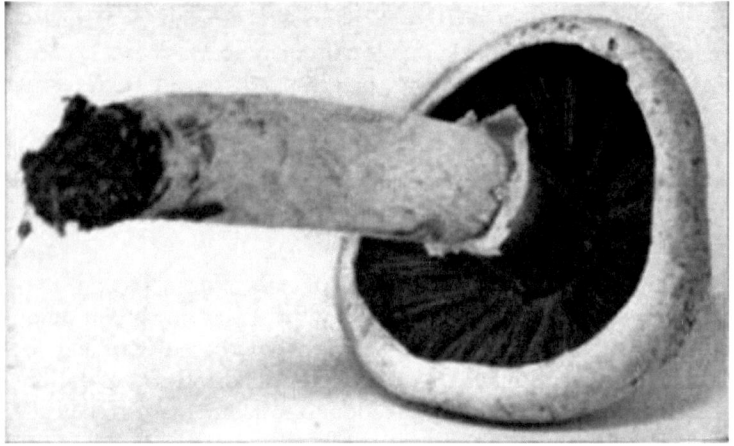

Figure. 1.—Agaricus campestris. View of under side showing stem, annulus, gills, and margin of pileus. (Natural size.)

The Stem or Stipe.—The stem in this plant, as in many other kinds, is attached to the pileus in the center. The purpose of the stem seems quite surely to be that of lifting the cap and the gills up above the ground, so that the spores can float in the currents of air and be readily scattered. The stem varies in length from 2–10 cm. and is about 1-1/2 cm. in diameter. It is cylindrical in form, and even, quite firm and compact, though sometimes there is a central core where the threads are looser. The stem is also white and fleshy, and is usually smooth.

The Ring.—There is usually present in the mature plant of *Agaricus campestris* a thin collar (*annulus*) or ring around the upper end of the stem. It is not a movable ring, but is joined to the stem. It is very delicate, easily rubbed off, or may be even washed off during rains. [Pg 3]

Parts Present in Other Mushrooms—The Volva.—Some other mushrooms, like the *deadly Amanita* (*Amanita phalloides*) and other species of the genus *Amanita*, have, in addition to the cap, gills, stem, and ring, a more or less well formed cup-like structure attached to the lower end of the stem, and from which the stem appears to spring. (Figs. 55, 72, etc.) This is the *volva*, sometimes popularly called the "death cup," or "poison cup." This structure is a very important one to observe, though its presence by no means indicates in all cases that the plant is poisonous. It will be described more in detail in treating of the genus *Amanita*, where the illustrations should also be consulted.

Figure 2.—Agaricus campestris. "Buttons" just appearing through the sod. Some spawn at the left lower corner. Soil removed from the front. (Natural size.)

Presence or Absence of Ring or Volva.—Of the mushrooms which have stems there are four types with respect to the presence or absence of the ring and volva. In the first type both the ring and volva are absent, as in the common fairy ring mushroom, *Marasmius oreades*; in the genus *Lactarius, Russula, Tricholoma, Clitocybe*, and others. In the second type the ring is present while the volva is absent, as in the common mushroom, *Agaricus campestris*, and its close allies; in the genus *Lepiota, Armillaria*, and others. In the third type the volva is present, but the ring is absent, as in the genus *Volvaria*, or *Amanitopsis*. In the fourth type both the ring and volva are present, as in the genus *Amanita*.

The Stem is Absent in Some Mushrooms.—There are also quite a large number of mushrooms which lack a stem. These usually grow on stumps, logs, or tree trunks, etc., and one side of the cap is attached directly to the wood on which the fungus is growing. [Pg 4] The pileus in such cases is lateral and shelving, that is, it stands out more or less like a shelf from the trunk or log, or in other cases is spread out flat on the surface of the wood. The shelving form is well shown in the beautiful *Claudopus nidulans*, sometimes called *Pleurotus nidulans*, and in other species of the genus *Pleurotus*, *Crepidotus*, etc. These plants will be described later, and no further description of the peculiarities in form of the mushrooms will be now attempted, since these will be best dealt with when discussing species fully under their appropriate genus. But the brief general description of form given above will be found useful merely as an introduction to the more detailed treatment. Chapter XXI should also be studied. For those who wish the use of a glossary, one is appended at the close of the book, dealing only with the more technical terms employed here.

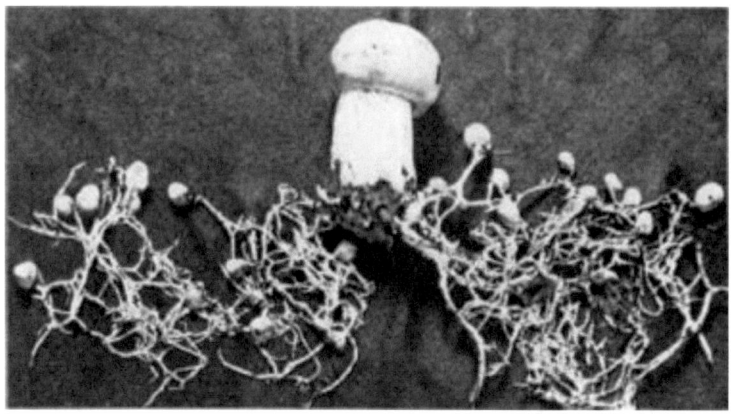

Figure 3.—Agaricus campestris. Soil washed from the "spawn" and "buttons," showing the young "buttons" attached to the strands of mycelium. (1-1/4 natural size.)

CHAPTER II. [Pg 5]

DEVELOPMENT OF THE MUSHROOM.

When the stems of the mushrooms are pulled or dug from the ground, white strands are often clinging to the lower end. These strands are often seen by removing some of the earth from the young plant, as shown in Fig. 2. This is known among gardeners as "spawn." It is through the growth and increase of this spawn that gardeners propagate the cultivated mushroom. Fine specimens of the spawn of the cultivated mushroom can be seen by digging up from a bed a group of very young plants, such a group as is shown in Fig. 3. Here the white strands are more numerous than can readily be found in the lawns and pastures where the plant grows in the feral state.

Figure 4.—Agaricus campestris. Sections of "buttons" at different stages, showing formation of gills and veil covering them. (Natural size.)

Nature of Mushroom Spawn.—This spawn, it should be clearly understood, is not spawn in the sense in which that word is used in fish culture; though it may be employed so readily in propagation of mushrooms. The spawn is nothing more than the vegetative portion of the plant. It is made up of countless numbers of delicate, tiny, white, jointed threads, the *mycelium*.

Mycelium of a Mold.—A good example of mycelium which is familiar to nearly every one occurs in the form of a white mold on bread or [Pg 6] on vegetables. One of the molds, so common on bread, forms at first a white cottony mass of loosely interwoven threads. Later the mold becomes black in color because of numerous small fruit cases containing dark spores. This last stage is the fruiting stage of the mold. The earlier stage is the growing, or vegetative, stage. The white mycelium threads grow in the bread and absorb food substances for the mold.

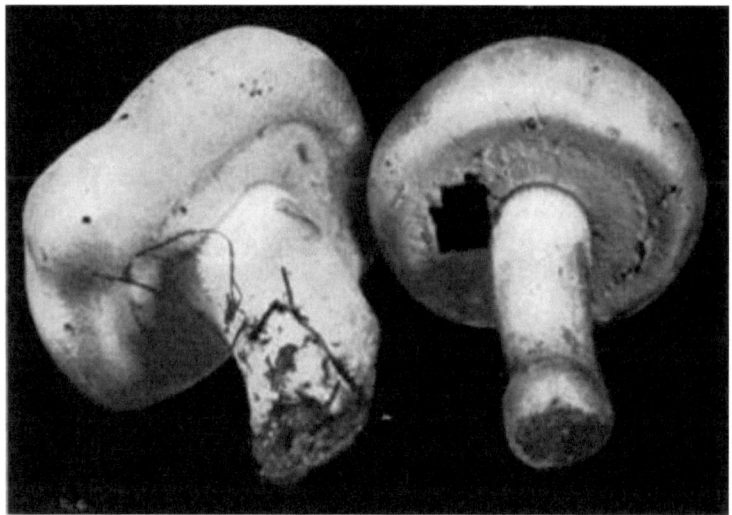

Figure 5.—Agaricus campestris. Nearly mature plants, showing veil stretched across gill cavity. (Natural size.)

Mushroom Spawn is in the Form of Strands of Mycelium.— Now in the mushrooms the threads of mycelium are usually inter-laced into definite strands or cords, especially when the mycelium is well developed. In some species these strands become very long, and are dark brown in color. Each thread of mycelium grows, or increases in length, at the end. Each one of the threads grows inde-pendently, though all are intertwined in the strand. In this way the strand of mycelium increases in length. It even branches as it ex-tends itself through the soil.

The Button Stage of the Mushroom.—The "spawn" stage, or strands of mycelium, is the vegetative or growing stage of the mushroom. These strands grow through the substance on which the fungus feeds. When the fruiting stage, or the mushroom, begins there appear small knobs or enlargements on these strands, and these are the beginnings of the button stage, as it is properly called. These knobs or young buttons are well shown in Fig. 3. They [Pg 7] begin by the threads of mycelium growing in great numbers out from the side of the cords. These enlarge and elongate and make their way toward the surface of the ground. They are at first very minute and grow from the size of a pinhead to that of a pea, and larger. Now they begin to elongate somewhat and the end enlarges as shown in the larger button in the figure. Here the two main parts of the mushroom are outlined, the stem and the cap. At this stage also the other parts of the mushroom begin to be outlined. The gills appear on the under side of this enlargement at the end of the button, next the stem. They form by the growth of fungus threads downward in radiating lines which correspond in position to the position of the gills. At the same time a veil is formed over the gills by threads which grow from the stem upward to the side of the button, and from the side of the button down toward the stem to meet them. This covers the gills up at an early period.

Figure 6. — Agaricus campestris. Under view of two plants just after rupture of the veil, fragments of the latter clinging both to margin of the pileus and to stem. (Natural size.)

From the Button Stage to the Mushroom. — If we split several of the buttons of different sizes down through the middle, we shall be able to see the position of the gills covered by the veil during their formation. These stages are illustrated in Fig. 4.

As the cap grows in size the gills elongate, and the veil becomes broader. But when the plant is nearly grown the veil ceases to grow, and then the expanding cap pulls so strongly on it that it is torn. Figure 5 shows the veil in a stretched condition just before it is ruptured, [Pg 8] and in Fig. 6 the veil has just been torn apart. The veil of the common mushroom is very delicate and fragile, as the illustration shows, and when it is ruptured it often breaks irregularly, sometimes portions of it clinging to the margin of the cap and portions clinging to the stem, or all of it may cling to the cap at times; but usually most of it remains clinging for a short while on the stem. Here it forms the annulus or ring.

Figure 7.—Agaricus campestris. Plant in natural position just after rupture of veil, showing tendency to double annulus on the stem. Portions of the veil also dripping from margin of pileus. (Natural size.)

The Color of the Gills.—The color of the gills of the common mushroom varies in different stages of development. When very young the gills are white. But very soon the gills become pink in color, and during the button stage if the veil is broken this pink color is usually present unless the button is very small. The pink color soon changes to dark brown after the veil becomes ruptured, and when the plants are quite old they are nearly black. This dark color of the gills is due to the dark color of the spores, which are formed in such great numbers on the surface of the gills.

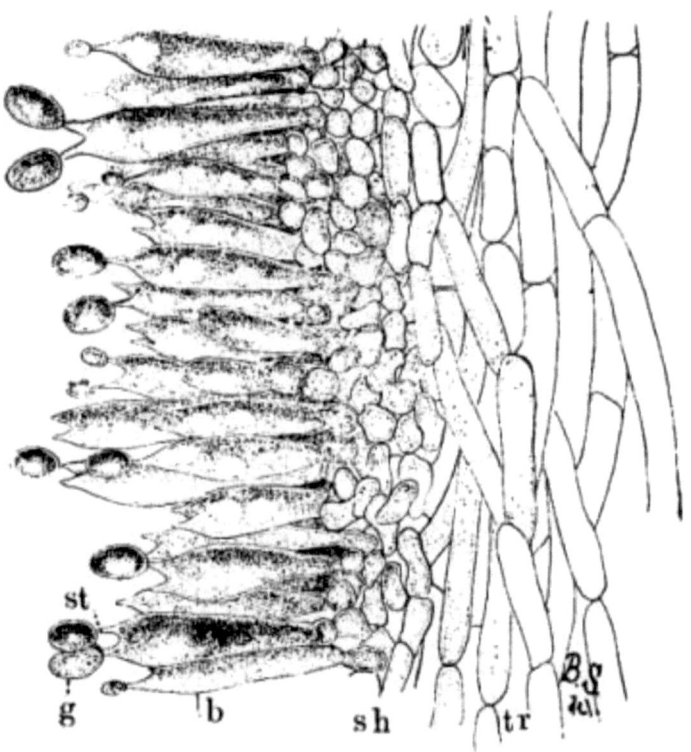

Figure 8.—Agaricus campestris. Section of gill showing *tr*==trama; *sh*==sub-hymenium; *b*==basidium, the basidia make up the hymenium; *st*==sterigma; *g*==spore. (Magnified.)

Structure of a Gill. — In Fig. 8 is shown a portion of a section across one of the gills, and it is easy to see in what manner the spores are borne. The gill is made up, as the illustration [Pg 9] shows, of mycelium threads. The center of the gill is called the *trama*. The trama in the case of this plant is made up of threads with rather long cells. Toward the outside of the trama the cells branch into short cells, which make a thin layer. This forms the *sub-hymenium*. The sub-hymenium in turn gives rise to long club-shaped cells which stand parallel to each other at right angles to the surface of the gill. The entire surface of the gill is covered with these club-shaped cells called *basidia* (sing. *basidium*). Each of these club-shaped cells bears either two or four spinous processes called *sterígmata* (sing. *sterígma*), and these in turn each bear a spore. All these points are well shown in Fig. 8. The basidia together make up the *hymenium*.

Figure 9.—Polyporus borealis, showing wound at base of hemlock spruce caused by falling tree. Bracket fruit form of Polyporus borealis growing from wound. (1/15 natural size.)

Wood Destroying Fungi.—Many of the mushrooms, and their kind, grow on wood. A visit to the damp forest during the summer months, or during the autumn, will reveal large numbers of these plants growing on logs, stumps, from buried roots or rotten wood, on standing dead trunks, or even on living trees. In the latter case the mushroom usually grows from some knothole or wound in the tree (Fig. 9). Many of the forms which appear on the trunks of dead or living trees are plants of tough or woody consistency. They are known as shelving or bracket fungi, or popularly as "fungoids" or "fungos." Both these latter words are very unfortunate and inappropriate. Many of these shelving or bracket fungi are perennial [Pg 10] and live from year to year. They may therefore be found during the winter as well as in the summer. The writer has found specimens over eighty years old. The shelves or brackets are the fruit bodies, and consist of the pileus with the fruiting surface below. The fruiting surface is either in the form of gills like *Agaricus*, or it is honeycombed, or spinous, or entirely smooth.

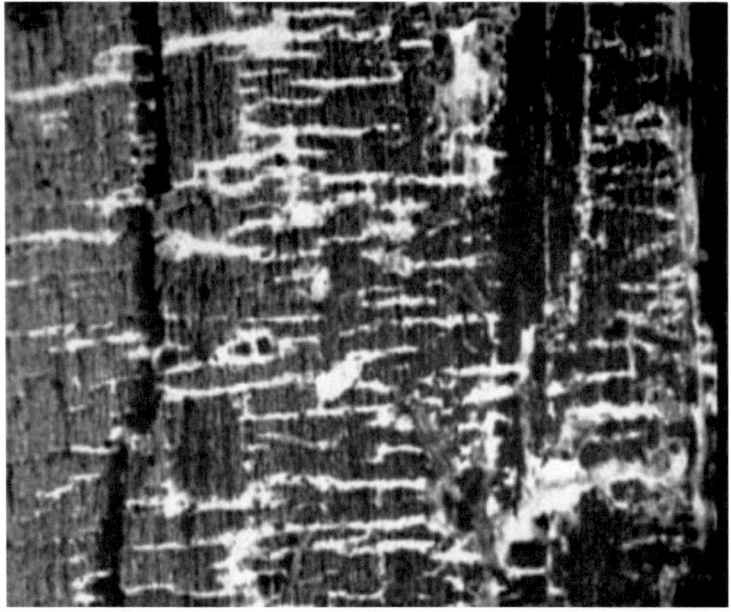

Figure 10.—Polyporus borealis. Strands of mycelium extending radially in the wood of the same living hemlock spruce shown in Fig. 9. (Natural size.)

Mycelium of the Wood Destroying Fungi.—While the fruit bodies are on the outside of the trunk, the mycelium, or vegetative part of the fungus, is within the wood or bark. By stripping off the bark from decaying logs where these fungi are growing, the mycelium is often found in great abundance. By tearing open the rotting wood it can be traced all through the decaying parts. In fact, the mycelium is largely if not wholly responsible for the rapid disintegration of the wood. In living trees the mycelium of certain bracket fungi enters through a wound and grows into the heart wood. Now the heart wood is dead and cannot long resist the entrance and destructive action of the mycelium. The mycelium spreads through the heart of the tree, causing it to rot (Fig. 10). When it has spread over a large feeding area it can then grow out through a wound or old knothole [Pg 11] and form the bracket fruit body, in case the knothole or wound has not completely healed over so as to imprison the fungus mycelium.

Plate 2, Figure 11.—Mycelium of Agaricus melleus on large door in passage coal mine, Wilkesbarre, Pa. (1/20 natural size.)

Fungi in Abandoned Coal Mines.—Mushrooms and bracket fungi grow in great profusion on the wood props or doors in abandoned coal mines, cement mines, etc. There is here an abundance of moisture, and the temperature conditions are more equable the year around. The conditions of environment then are very favorable for the rapid growth of these plants. They develop in midwinter as well as in summer.

Mycelium of Coal Mine Fungi. — The mycelium of the mushrooms and bracket fungi grows in wonderful profusion in these abandoned coal mines. So far down in the moist earth the air in the tunnels or passages where the coal or rock has been removed is at all times nearly saturated with moisture. This abundance of moisture, with the favorable temperature, permits the mycelium to grow on the surface of the wood structures as readily as within the wood.

In the forest, while the air is damp at times, it soon dries out to such a degree that the mycelium can not exist to any great extent on the outer surface of the trunks and stumps, for it needs a great percentage of moisture for growth. The moisture, however, is abundant within the stumps or tree trunks, and the mycelium develops abundantly there.

So one can understand how it is that deep down in these abandoned mines the mycelium grows profusely on the surface of doors and wood props. Figure 11 is from a flashlight photograph, taken by the writer, of a beautiful growth on the surface of one of the doors in an abandoned coal mine at Wilkesbarre, Pa., during September, 1896. The specimen covered an area eight by ten feet on the surface of the door. The illustration shows very well the habit of growth of the mycelium. At the right is the advancing zone of growth, marked by several fan-shaped areas. At the extreme edge of growth the mycelium presents a delicate fringe of the growing ends where the threads are interlaced uniformly over the entire area. But a little distance back from the edge, where the mycelium is older, the threads are growing in a different way. They are now uniting into definite strands. Still further back and covering the larger part of the sheet of mycelium lying on the surface of the door, are numerous long, delicate tassels hanging downward. These were formed by the attempt on the part of the mycelium at numerous places to develop strands at right angles to the surface of the door. There being nothing to support them in their attempted aerial flight, [Pg 12] they dangle downward in exquisite fashion. The mycelium in this condition is very soft and perishable. It disappears almost at touch.

On the posts or wood props used to support the rock roof above, the mycelium grows in great profusion also, often covering them with a thick white mantle, or draping them with a fabric of elegant texture. From the upper ends of the props it spreads out over the rock roof above for several feet in circumference, and beautiful white pendulous tassels remind one of stalactites.

Figure 12. — Agaricus campestris. Spore print. (Natural size.)

Direction in Growth of Mushrooms. — The direction of growth which these fungi take forms an interesting question for study. The common mushroom, the *Agaricus*, the amanitas, and other central stemmed species grow usually in an upright fashion; that is, the stem is erect. The cap then, when it expands, stands so that it is parallel with the surface of the earth. Where the cap does not fully expand, as in the campanulate forms, the pileus is still oriented horizontally, that is, with the gills downward. Even in such species, where the stems are ascending, the upper end of the stem curves so that the cap occupies the usual position with reference to the surface of the earth. This is beautifully shown in the case of those plants which grow on the side of trunks or stumps, where the stems could not well grow directly upward without hugging close to the side of the trunk, and then there would not be room for the expansion of the cap. This is well shown in a number of species of *Mycena*.

In those species where the stem is sub-central, i. e., set toward one side of the pileus, or where it is definitely lateral, the pileus is also expanded in a horizontal direction. From these lateral stemmed species there is an easy transition to the stemless forms which are [Pg 13] sessile, that is, the shelving forms where the pileus is itself attached to the trunk, or other object of support on which it grows.

Where there is such uniformity in the position of a member or part of a plant under a variety of conditions, it is an indication that there is some underlying cause, and also, what is more important, that this position serves some useful purpose in the life and well being of the plant. We may cut the stem of a mushroom, say of the *Agaricus campestris*, close to the cap, and place the latter, gills downward, on a piece of white paper. It should now be covered securely with a small bell jar, or other vessel, so that no currents of air can get underneath. In the course of a few hours myriads of the brown spores will have fallen from the surface of the gills, where they are borne. They will pile up in long lines along on either side of all the gills and so give us an impression, or spore print, of the arrangement of the gills on the under side of the cap as shown in Fig. 12. A white spore print from the smooth lepiota (*L. naucina*) is shown in Fig. 13. This horizontal position of the cap then favors the falling of the spores, so that currents of air can scatter them and aid in the distribution of the fungus.

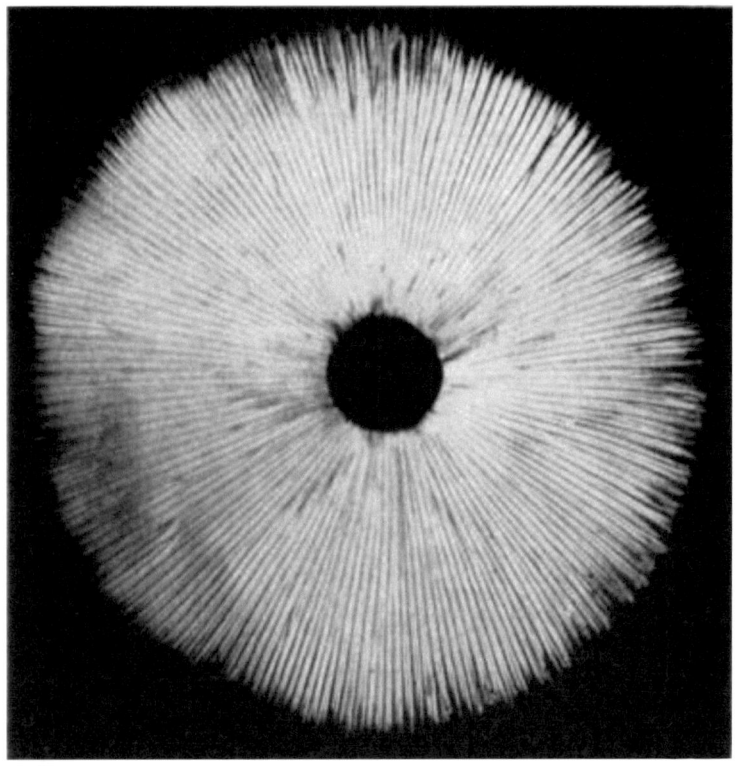

Figure 13.—Lepiota naucina. Spore print. (Natural size.)

But some may enquire how we know that there is any design in the horizontal position of the cap, and that there is some cause which brings about this uniformity of position with such entire harmony among such dissimilar forms. When a mushroom with a comparatively long stem, not quite fully matured or expanded, is pulled and laid on its side, or held in a horizontal position for a time, the upper part of the stem where growth is still taking place will curve upward [Pg 14] so that the pileus is again brought more or less in a horizontal position.

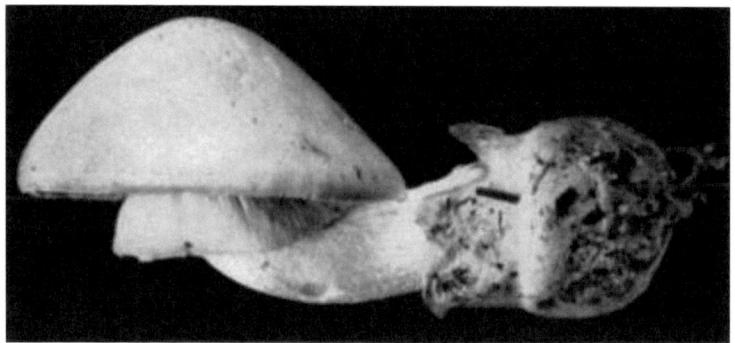

Figure 14.—Amanita phalloides. Plant turned to one side by directive force of gravity, after having been placed in a horizontal position. (Natural size.)

In collecting these plants they are often placed on their side in the collecting basket, or on a table when in the study. In a few hours the younger, long stemmed ones have turned upward again. The plant shown in Fig. 14 (*Amanita phalloides*) was placed on its side in a basket for about an hour. At the end of the hour it had not turned. It was then stood upright in a glass, and in the course of a few hours had turned nearly at right angles. The stimulus it received while lying in a horizontal position for only an hour was sufficient to produce the change in direction of growth even after the upright position had been restored. This is often the case. Some of the more sensitive of the slender species are disturbed if they lie for only ten or fifteen minutes on the side. It is necessary, therefore, when collecting, if one wishes to keep the plants in the natural position for photographing, to support them in an upright position when they are being carried home from the woods.

The cause of this turning of the stem from the horizontal position, so that the pileus will be brought parallel with the surface of the earth, is the stimulus from the force of gravity, which has been well demonstrated in the case of the higher plants. That is, the force which causes the stems of the higher plants to grow upward also regulates the position of the cap of the pileated fungi. The reason for this is to be seen in the perfection with which the spores are shed from the surfaces of the gills by falling downward and out from the crevices between. The same is true with the shelving fungi on trees,

[Pg 15] etc., where the spores readily fall out from the pores of the honey-combed surface or from between the teeth of those sorts with a spiny under surface. If the caps were so arranged that the fruiting surface came to be on the upper side, the larger number of the spores would lodge in the crevices between the extensions of the fruiting surface. Singularly, this position of the fruiting surface does occur in the case of one genus with a few small species.

Interesting examples of the operation of this law are sometimes met with in abandoned coal mines, or more frequently in the woods. In abandoned mines the mushrooms sometimes grow from the mycelium which spreads out on the rock roof overhead. The rock roof prevents the plant from growing upright, and in growing laterally the weight of the plant together with the slight hold it can obtain on the solid rock causes it to hang downward. The end of the stem then curves upward so that the pileus is brought in a horizontal position. I have seen this in the case of *Coprinus micaceus* several times.

Figure 15.—Polyporus applanatus. From this view the larger cap is in the normal position in which it grew on the standing tree. Turn one fourth way round to the right for position of the plant after the tree fell. (1/6 natural size.)

In the woods, especially in the case of the perennial shelving fungi, interesting cases are met with. Figure 15 illustrates one of these peculiar forms of *Polyporus (Fomes) applanatus*. This is the species so often collected as a "curio," and on account of its very white under surface is much used for etching various figures. In the figure the larger cap which is horizontal represents the position of the plant when on the standing maple trunk. When the tree fell [Pg 16] the shelf was brought into a perpendicular position. The fungus continued to grow, but its substance being hard and woody it cannot turn as the mushroom can. Instead, it now grows in such a way as to form several new caps, all horizontal, i. e., parallel with the surface of the earth, but perpendicular to the old shelf. If the page is turned one-fourth way round the figure will be brought in the position of the plant when it was growing on the fallen log.

Plate 3, Figure 16.—Dædalea ambigua. Upper right-hand shows normal plant in normal position when on tree. Upper left-hand shows abnormal plant with the large cap in normal position when growing on standing tree. Lower plant shows same plant in position after the tree fell, with new caps growing out in horizontal direction. (Lower plant 1/2 natural size.)

[Pg 17] Another very interesting case is shown in the ambiguous trametes (*Trametes ambigua*), a white shelving fungus which occurs in the Southern States. It is shown in Fig. 16. At the upper right hand is shown the normal plant in the normal position. At the upper left hand is shown an abnormal one with the large and first formed cap also in the normal position as it grew when the tree was standing. When the tree fell the shelf was on the upper side of the log. Now numerous new caps grew out from the edge as shown in the lower figure, forming a series of steps, as it were, up one side and down the other.

CHAPTER III.

GILL BEARING FUNGI: AGARICACEAE. [A]

The gill bearing fungi are known under the family *Agaricaceæ*, or popularly the agarics. They are distinguished by the fruiting area being distributed over the surface of plate-like or knife-like extensions or folds, usually from the under surface of the cap. These are known as the gills, or lamellæ, and they usually radiate from a common point, as from or near the stem, when the stem is present; or from the point of attachment of the pileus when the stem is absent. The plants vary widely in form and consistency, some being very soft and soon decaying, others turning into an inky fluid, others being tough and leathery, and some more or less woody or corky. The spores when seen in mass possess certain colors, white, rosy, brown or purple brown, black or ochraceous. While a more natural division of the agarics can be made on the basis of structure and consistency, the treatment here followed is based on the color of the spores, the method in vogue with the older botanists. While this method is more artificial, it is believed to be better for the beginner, especially for a popular treatment. The sections will be treated in the following order:

1. The purple-brown-spored agarics.
2. The black-spored agarics.
3. The white-spored agarics.
4. The rosy-spored agarics.
5. The ochre-spored agarics.

FOOTNOTES:

[A] For analytical keys to the families and genera see Chapter XXIV .

[Pg 18] CHAPTER IV.

THE PURPLE-BROWN-SPORED AGARICS. [B]

The members of this subdivision are recognized at maturity by the purple-brown, dark brown or nearly black spores when seen in mass. As they ripen on the surface of the gills the large number give the characteristic color to the lamellæ. Even on the gills the purple tinge of the brown spores can often be seen. The color is more satisfactorily obtained when the spores are caught in mass by placing the cap, gills downward, on white paper.

AGARICUS Linn. (PSALLIOTA Fr.)

In the genus *Agaricus* the spores at maturity are either purple-brown in mass or blackish with a purple tinge. The annulus is present on the stem, though disappearing soon in some species, and the stem is easily separated from the substance of the pileus. The gills are free from the stem, or only slightly adnexed. The genus is closely related to *Stropharia* and the species of the two genera are by some united under one genus (*Psalliota*, Hennings). Peck, 36th Report, N. Y. State Mus., p. 41–49, describes 7 species. Lloyd Mycol. Notes, No. 4, describes 8 species. C. O. Smith, Rhodora, I: 161–164, 1899, describes 8 species.

Agaricus (Psalliota) campestris Linn. **Edible.**—This plant has been quite fully described in the treatment of the parts of the mushroom, and a recapitulation will be sufficient here. It grows in lawns, pastures, by roadsides, and even in gardens and cultivated fields. A few specimens begin to appear in July, it is more plentiful in August, and abundantly so in September and October. It is 5–8 cm. high (2–3 inches), the cap is 5–12 cm. broad, and the stem 8–12 mm. in thickness.

The **pileus** is first rounded, then convex and more or less expanded. The surface at first is nearly smooth, presenting a soft, silky appearance from numerous loose fibrils. The surface is sometimes more or less torn into triangular scales, especially as the plants become old. The color is usually white, but varies more or less to light brown, especially in the scaly forms, where the scales may be [Pg 19] quite prominent and dark brown in color. Sometimes the color is brownish before the scales appear. The flesh is white. The **gills** in the young button stage are white. They soon become pink in color and after the cap is expanded they quickly become purple brown, dark brown, and nearly black from the large number of spores on their surfaces. The gills are free from the stem and rounded behind (near the stem). The **stem** is white, nearly cylindrical, or it tapers a little toward the lower end. The flesh is solid, though the central core is less firm. The **veil** is thin, white, silky, and very frail. It is stretched as the cap expands and finally torn so that it clings either as an annulus around the stem, or fragments cling around the margin of the cap. Since the **annulus** is so frail it shrivels as the plant ages and becomes quite inconspicuous or disappears entirely (see Figs. 1–7).

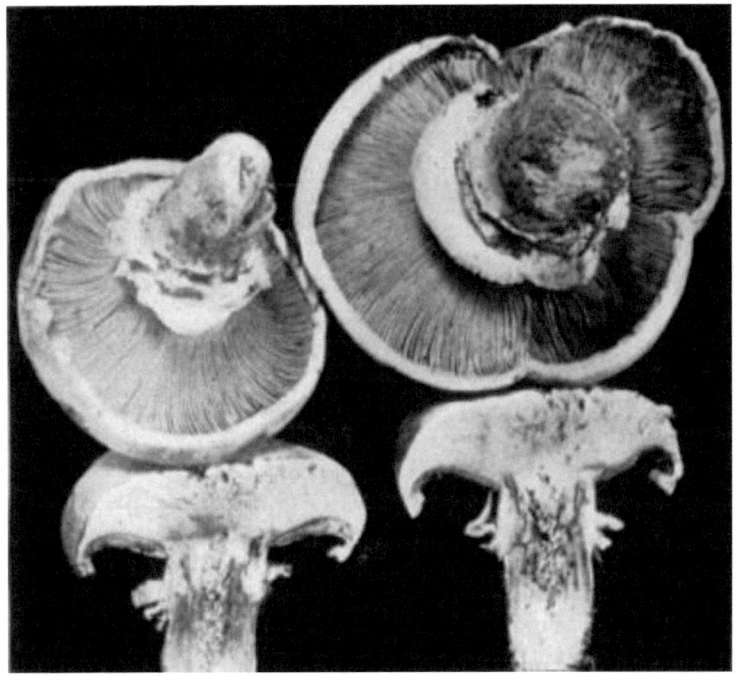

Figure 17.—Agaricus rodmani. Entirely white, showing double veil or ring. (Natural size.) Copyright.

Variations in the surface characters of the cap and stem have led some to recognize several varieties. This is known as the common mushroom and is more widely known and collected for food than any [Pg 20] other. It is also cultivated in mushroom houses, cellars, caves, abandoned mines, etc.

Agaricus (Psalliota) rodmani Pk. **Edible.**—Rodman's mushroom, *Agaricus rodmani*, grows in grassy places along streets of cities, either between the curbing and the walk, or between the curbing and the pavement. It is entirely white or whitish and sometimes tinged with yellowish at the center of the pileus. The plants are 4–8 cm. high, the cap 5–8 cm. broad and the stem 1–2 cm. in thickness.

Figure 18.—Agaricus arvensis, fairy ring.

The **pileus** is rounded, and then convex, very firm, compact and thick, with white flesh. The **gills** are crowded, first white, then pink, and in age blackish brown. The **stem** is very short, solid, nearly cylindrical, not bulbous. The **annulus** is quite characteristic, being very thick, with a short limb, and double, so that it often appears as two distinct rings on the middle or lower part of the stem as shown in Fig. 17. This form of the annulus is probably due to the fact that the thick part of the margin of the pileus during the young stage rests between the lower and upper part of the annulus, i. e., the thick veil is attached both to the inner and outer surface of the margin of the cap, and when it is freed by the expansion of the pileus it remains as a double ring. It is eagerly sought and much relished by several persons at Ithaca familiar with its edible qualities.

The plant closely resembles A. campestris var., edulis, Vittad. (See Plate 54, Bresadola, I Funghi Mangerecci e Velenosi, 1899) and is probably the same.

Figure 19.—Agaricus silvicola. White to cream color, or yellow stains. (Natural size.) Copyright.

[Pg 21] **Agaricus (Psalliota) arvensis** Schaeff. **Edible.**—The field mushroom, or horse mushroom, *Agaricus arvensis*, grows in fields or pastures, sometimes under trees and in borders of woods. One form is often white, or yellowish white, and often shows the yellow color when dried. The plant sometimes occurs in the form of a fairy ring as shown in Fig. 18. It is 5–12 cm. high, the cap from 5–15 cm. broad and the stem 8–15 mm. in thickness.

The **pileus** is smooth, quite thick and firm, convex to expanded. The **gills** are first white, then tinged with pink and finally blackish brown. The **stem** is stout, nearly cylindrical, hollow, bulbous. The veil is double like that of *Agaricus placomyces*, the upper or inner [Pg 22] layer remaining as a membrane, while the lower or outer layer is split radially and remains in large patches on the lower surface of the upper membrane.

Figure 20. — Agaricus silvicola, showing radiately torn lower part of veil. (Natural size.) Copyright.

Agaricus (Psalliota) silvicola Vittad. **Edible.** — The *Agaricus silvicola* grows in woods, groves, etc., on the ground, and has been found also in a newly made garden in the vicinity of trees near the woods. It is an attractive plant because of its graceful habit and the delicate shades of yellow and white. It ranges from 10–20 cm. high, the cap is 5–12 cm. broad and the stem 6–10 mm. in thickness.

The **pileus** becomes convex, and expanded or nearly flat, and often with an elevation or umbo in the center. It is thin, smooth, whitish and often tinged more or less deeply with yellow (sulfur or ochraceous) and is sometimes tinged with pink in the center. The flesh is whitish or tinged with pink. The **gills** when very young are whitish, then pink, and finally dark brown or blackish brown, much crowded, and distant from the stem. The **stem** is long, nearly cylindrical, whitish, abruptly enlarged below into a bulb. It is often yellowish below, and especially in drying becomes stained with yellow. The **ring** is thin, membranaceous, delicate, sometimes with broad, soft, floccose patches on the under side. The ring usually appears single, but sometimes the **veil** is seen to be double, and the outer or lower portion tends to split radially as in *A. arvensis* or *A. placomyces*. This is well shown in large specimens, and especially as the veil is stretched over the gills as shown in Fig. 20.

From the form of the plant as well as the peculiarities of the veil in the larger specimens, it is related to *A. arvensis* and *A. placomyces*, more closely to the former. It occurs during mid-summer and early autumn. Figure 10 is from plants (No. 1986 C. U. herbarium) collected in open woods at Ithaca.

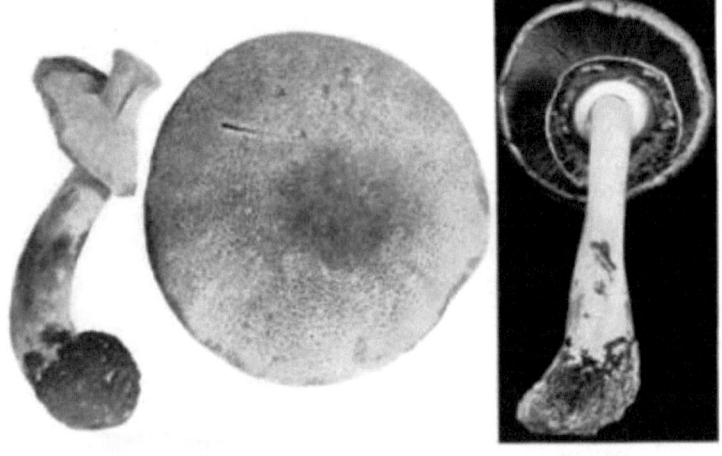

FIGURE 21.   FIGURE 22.

Plate 4.—Agaricus placomyces. Figure 21.—Upper view of cap, side view of stem. Figure 22.—Under view of plant showing radiately torn under side of the double veil. (3/4 natural size.) Copyright.

[Pg 23]

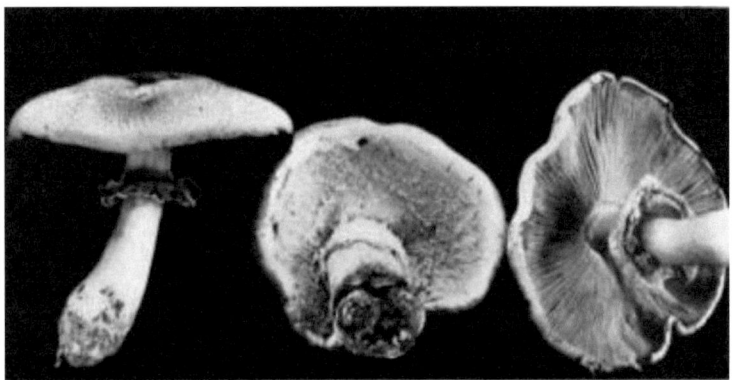

Plate 5, Figure 23.—Agaricus placomyces. Three different views, see text for explanations. Dark scales on cap. (Natural size.) Copyright.

Agaricus (Psalliota) subrufescens Pk. **Edible.**—The *Agaricus subrufescens* was described by Dr. Peck from specimens collected on a compost heap composed chiefly of leaves, at Glen Cove, Long Island. It occurs sometimes in greenhouses. In one case reported by Peck it appeared in soil prepared for forcing cucumbers in a greenhouse in Washington, D. C.

According to the description the **pileus** becomes convex or broadly expanded, is covered with silky hairs and numerous minute scales. The color is whitish, grayish or dull reddish brown, the center being usually smooth and darker, while the flesh is white. The **gills** change from white to pinkish and blackish brown in age. The **stem** is long, nearly cylindrical or somewhat enlarged or bulbous at the base, first stuffed, then hollow, white. The **annulus** is thick, and the under side marked by loose threads or scales.

This plant is said to differ from the common mushroom (*A. campestris*) in the more deeply hemispherical cap of the young plant, the

42

hollow and somewhat bulbous stem, and in the scales on the under side of the annulus. In fresh plants the flesh has also a flavor of almonds. It is closely related to **A. silvaticus** Schaeff., p. 62, T. 242, Icones Fung. Bav. etc., 1770, if not identical with it. *A. silvaticus* has light ochraceous or subrufescent scales on the cap, a strong odor, and occurs in gardens as well as in the woods.

Agaricus (Psalliota) fabaceus Berk., was described in Hooker's London Journal of Botany, **6**: 314, 1847, from specimens collected in Ohio. The plant is white and is said to have a strong but not unpleasant odor. *Agaricus amygdalinus* Curt., from North Carolina, and of which no description was published, was so named on account of the almond-like flavor of the plant. Dr. Farlow suggests (Proc. Bost. Soc. Nat. Hist. **26**: 356–358, 1894) that *A. fabaceus, amygdalinus*, and *subrufescens* are identical.

Agaricus (Psalliota) placomyces Pk. **Edible.** — The flat-cap mushroom, *Agaricus placomyces* Pk., occurs in borders of woods or under trees from June to September. According to Peck it occurs in borders of hemlock woods, or under hemlock trees. At Ithaca it is not always associated with hemlock trees. The largest specimens found here were in the border of mixed woods where hemlock was a constituent. It has been found near and under white pine trees in lawns, around the Norway spruce and under the Norway spruce. The plants are from 5–15 cm. high, the cap from 5–12 cm. in diameter, and the stem 6–8 mm. in thickness. [Pg 24]

The **pileus** when young is broadly ovate, then becomes convex or fully expanded and flat in age, and is quite thin. The ground color is whitish, often with a yellowish tinge, while the surface is ornamented with numerous minute brownish scales which are scattered over a large part of the cap, but crowded or conjoined at the center into a large circular patch. This gives to the plant with its shapely form a beautiful appearance. In the young stage the entire surface of the pileus is quite evenly brown. As it expands the outer brown portion is torn asunder into numerous scales because the surface threads composing this brown layer cease to grow. These scales are farther apart toward the margin of the cap, because this portion of the cap always expands more than the center, in all mushrooms. The **gills** are at first white, or very soon pink in color, and in age are blackish brown. Spores 5–8 × 3–4 μ.

The **stem** is nearly cylindrical, hollow or stuffed, white or whitish, smooth, bulbous, and the bulb is sometimes tinged with yellow. The **veil** is very handsome, and the way in which the annulus is formed from it is very interesting. The veil is quite broad, and it is double, that is, it consists of two layers which are loosely joined by threads. In the young stage the veil lies between the gills and the lower two-thirds of the stem. As the pileus expands the lower (outer part) layer of the veil is torn, often in quite regular radiating portions, as shown in Fig. 22. An interesting condition of the veil is shown in the middle plant in Fig. 23. Here the outer or lower layer of the veil did not split radially, but remained as a tube surrounding the stem, while the two layers were separated, the inner one being still stretched over the gills. It is customary to speak of the lower part of the veil as the outer part when the cap is expanded and the veil is still stretched across over the gills, while the upper portion is spoken of as the inner layer or part. It is closely related to *A. arvensis*, and may represent a wood inhabiting variety of that species.

Agaricus (Psalliota) comtulus Fr. — This pretty little agaric seems to be rather rare. It was found sparingly on several occasions in open woods under pines at Ithaca, N. Y., during October, 1898. Lloyd reports it from Ohio (Mycolog. Notes, No. 56, Nov. 1899), and Smith from Vermont (Rhodora I, 1899). Fries' description (Epicrisis, No. 877) runs as follows: "Pileus slightly fleshy, convex, plane, obtuse, nearly smooth, with appressed silky hairs, stem hollow, subattenuate, smooth, white to yellowish, annulus fugacious; gills free, crowded, broad in front, from flesh to rose color. In damp grassy places. Stem 2 inches by 2 lines, at first floccose stuffed. Pileus 1–1-1/2 inch diameter. Color from white to yellowish."

Figure 24.—Agaricus comtulus (natural size, sometimes larger). Cap creamy white with egg-yellow stains, smoky when older. Stem same color; gills grayish, then rose, then purple brown. Copyright.

[Pg 25] The plants collected at Ithaca are illustrated in Fig. 24 from a photograph of plants (No. 2879 C. U. herbarium). My notes on these specimens run as follows: Plant 3–6 cm. high, pileus 1.5–3 cm. broad, stem 3–4 mm. in thickness. **Pileus** convex to expanded, fleshy, thin on the margin, margin at first incurved, creamy white with egg yellow stains, darker on the center, in age somewhat darker to umber or fuliginous, moist when fresh, surface soon dry, flesh tinged with yellow. The **gills** are white when young, then grayish to pale rose, and finally light purple brown, rounded in front, tapering behind (next the stem) and rounded, free from the stem, 4–5 mm. broad. **Basidia** clavate, 25–30 × 5–6 μ. **Spores** small, oval, 3–4 × 2–3 μ, in mass light purple brown. The **stem** tapers above, is sub-bulbous below, yellowish and stained with darker yellowish threads below the annulus, hollow, fibrous, fleshy. The **veil** whitish stained with yellow, delicate, rupturing irregularly, portions of it clinging to margin of the pileus and portions forming a delicate ring. When parts of the plant come in contact with white paper a blue stain is apt to be imparted to the paper, resembling the reaction of iodine on starch. This peculiarity has been observed also in the case of another species of *Agaricus*. The species is regarded with suspicion by some. I collected the plant also at Blowing Rock, N. C., in September, 1899. The caps of these specimens measure 4 cm. in diameter.

Agaricus diminutivus Pk., is a closely related species. It is distinguished chiefly by its somewhat larger size, and purplish to reddish brown hairs on the surface of the pileus, and by the somewhat larger spores, which, however, are small. I have found it at Ithaca, the surface of the pileus hairy, with beautiful, triangular, soft, appressed, purplish scales.

HYPHOLOMA Fr.

In the genus *Hypholoma* the spores are purple brown, the gills attached to the stem, and the veil when ruptured clings to the margin of the cap instead of to the stem, so that a ring is not formed, or only rarely in some specimens. The stem is said to be continuous with the substance of the cap, that is, it is not easily separated from it. The genus is closely related to *Agaricus (Psalliota)* and *Stropharia*, from both of which it differs in the veil not forming a ring, but clinging to the margin of the cap. It further differs from *Agaricus* in the stem being continuous with the substance of the cap, while *Stropharia* seems to differ in this respect in different species. The plants grow both on the ground and on wood. There are several species which are edible and are very common. Peck gives a synopsis of six species in the 49th Report New York State Mus., page 61, 1896, and Morgan describes 7 species in Jour. Cinn. Soc. Nat. Hist. **6**: 113–115.

Hypholoma sublateritium Schaeff. **Edible**, *bitter sometimes*. The name of this species is derived from the color of the cap, which is nearly a brick red color, sometimes tawny. The margin is lighter in color. The plants grow usually in large clusters on old stumps or frequently appearing on the ground from buried portions of stumps or from roots. There are from six to ten, or twenty or more plants in a single cluster. A single plant is from 8–12 cm. high, the cap is 5–8 cm. broad, and the stem 6–8 mm. in thickness.

The **pileus** is convex to expanded, smooth, or sometimes with loose threads from the veil, especially when young, even, dry. The flesh is firm, whitish, and in age becoming somewhat yellowish. The **gills** are adnate, sometimes decurrent by a little tooth, rather crowded, narrow, whitish, then dull yellow, and becoming dark from the spores, purplish to olivaceous. The **stem** usually tapers downward, is firm, stuffed, smooth, or with remnants of the veil giving it a floccose scaly appearance, usually ascending because of the crowded growth. The **veil** is thin and only manifested in the young stage of the plant as a loose weft of threads. As the cap expands the veil is torn and adheres to the margin, but soon disappears.

Plate 6, Figure 25.—Hypholoma sublateritium. Cap brick-red or tawny. (Natural size, often larger.) Copyright.

[Pg 27]

Plate 7, Figure 26.—Hypholoma appendiculatum (natural size, often larger). White floccose scales on cap (var. coroniferum) and appendiculate veil; caps whitish or brown, tawny, or tinge of ochre. Gills white, then purple-brown. Copyright.

The flesh of this plant is said by European writers to be bitter to the taste, and it is regarded there as poisonous. This character seems to be the only distinguishing one between the *Hypholoma sublateritium* Schaeff., of Europe, and the *Hypholoma perplexum* Pk., of this country which is edible, and probably is identical with *H. sublateritium*. If the plant in hand agrees with this description in other respects, and is not bitter, there should be no danger in its use. According to Bresadola, the bitter taste is not pronounced in *H. sublateritium*. The taste probably varies as it does in other plants. For example, in *Pholiota præcox*, an edible species, I detected a decided bitter taste in plants collected in June, 1900. Four other persons were requested to taste the plants. Two of them pronounced them bitter, while two did not detect the bitter taste.

There is a variety of *Hypholoma sublateritium*, with delicate floccose scales in concentric rows near the margin of the cap, called *var. squamosum* Cooke. This is the plant illustrated in Fig. 25, from specimens collected on rotting wood in the Cascadilla woods, Ithaca, N. Y. It occurs from spring to autumn.

Hypholoma epixanthum Fr., is near the former species, but has a yellow pileus, and the light yellow gills become gray, not purple.

Hypholoma appendiculatum Bull. **Edible.**—This species is common during late spring and in the summer. It grows on old stumps and logs, and often on the ground, especially where there are dead roots. It is scattered or clustered, but large tufts are not formed as in *H. sublateritium.* The plants are 6–8 cm. high, the cap 5–7 cm. broad, and the stem 4–6 mm. in thickness.

The **pileus** is ovate, convex to expanded, and often the margin elevated, and then the cap appears depressed. It is fleshy, thin, whitish or brown, tawny, or with a tinge of ochre, and becoming pale in age and when dry. As the plant becomes old the pileus often cracks in various ways, sometimes splitting radially into several lobes, and then in other cases cracking into irregular areas, showing the white flesh underneath. The surface of the pileus when young is sometimes sprinkled with whitish particles giving it a mealy appearance. The **gills** are attached to the stem, crowded, becoming more or less free by breaking away from the stem, especially in old plants. They are white, then flesh colored, brownish with a slight purple tinge. The **stem** is white, smooth, or with numerous small [Pg 28] white particles at the apex, becoming hollow. The **veil** is very delicate, white, and only seen in quite young plants when they are fresh. It clings to the margin of the cap for a short period, and then soon disappears.

Figure 27.—Hypholoma appendiculatum (natural size), showing appendiculate veil. Copyright.

Sometimes the pileus is covered with numerous white, delicate floccose scales, which give it a beautiful appearance, as in Fig. 26, from specimens (No. 3185 C. U. herbarium), collected on the campus of Cornell University among grass. The entire plant is very brittle, and easily broken. It is tender and excellent for food. I often eat the caps raw.

Hypholoma candolleanum Fr., occurs in woods on the ground, or on very rotten wood. It is not so fragile as *H. appendiculatum* and the gills are dark violaceous, not flesh color as they are in *H. appendiculatum* when they begin to turn, and nearly free from the stem.

Hypholoma lacrymabundum Fr.—This plant was found during September and October in wet grassy places in a shallow ditch by the roadside, and in borders of woods, Ithaca, N. Y., 1898. The plants are scattered or clustered, several often joined at the base of the stem. They are 4–8 cm. high, the cap 2–5 cm. broad, and the stem 4–8 mm. in thickness.

Figure 28.—Hypholoma lacrymabundum (natural size). Cap and stem tawny or light yellowish, with intermediate shades or shades of umber, surface with soft floccose scales. Copyright.

[Pg 29] The **pileus** is convex to expanded, sometimes broadly umbonate in age, and usually with radiating wrinkles extending irregularly. On the surface are silky or tomentose threads not much elevated from the surface, and as the plant ages these are drawn into triangular scales which are easily washed apart by the rains. The color is tawny or light yellowish with intermediate shades, darker on the umbo and becoming darker in age, sometimes umber colored, and stained with black, especially after rains where the spores are washed on the pileus. The flesh is tinged with light yellow, or tawny, or brown, soft, and easily broken. The **gills** are sinuate, adnate, somewhat ventricose, very rarely in abnormal specimens anastomosing near the margin of the pileus, at first light yellowish, then shading to umber and spotted with black and rusty brown as the spores mature, easily breaking away from the stipe, whitish on the edge. Drops of moisture sometimes are formed on the gills. **Basidia** abruptly clavate, 30–35 × 10–12 μ. **Cystidia** hyaline, thin walled, projecting above the hymenium 40 μ, and 14–15 μ

51

broad. Spores black, purple tinged, broadly elliptical and somewhat curved, 9–11 × 7–8 µ. [Pg 30]

The **stem** is fleshy to fibrous, the same color as the pileus, floccose scaly more or less up to the veil, smooth or white pruinose above the veil, straight or curved, somewhat striate below.

The **veil** in young plants is hairy, of the same texture as the surface of the pileus, torn and mostly clinging to the margin of the pileus, and disappearing with age.

The general habit and different stages of development as well as some of the characters of the plant are shown in Fig. 28 (No. 4620 Cornell University herbarium). The edible qualities of this plant have not been tested.

Hypholoma rugocephalum Atkinson.—This interesting species grows in damp places in woods. The plants are tufted or occur singly. They are 8–12 cm. high, the cap 6–10 cm. broad, and the stem 6–10 mm. in thickness.

The **pileus** is convex to expanded, and the margin at last revolute (upturned). The surface is marked by strong wrinkles (rugæ), which radiate irregularly from the center toward the margin. The pileus is broadly umbonate, fleshy at the center and thinner toward the margin, the flesh tinged with yellow, the surface slightly viscid, but not markedly so even when moist, smooth, not hairy or scaly, the thin margin extending little beyond ends of the gills. The color is tawny (near fulvus). The **gills** are adnate, slightly sinuate, 5–7 mm. broad, in age easily breaking away from the stem and then rounded at this end, spotted with the black spores, lighter on the edge. The **spores** are black in mass (with a suggestion of a purple tinge), oval to broadly elliptical, inequilateral, pointed at each end, echinulate, or minutely tuberculate, 8–11 × 6–8 µ. The **basidia** are short, cylindrical; **cystidia** cylindrical, somewhat enlarged at the free end, hyaline, delicate, thin-walled, in groups of two to six or more (perhaps this is partly responsible for the black spotted condition of the gills). The **stem** is cylindrical, even, somewhat bulbous, of the same color as the pileus, but lighter above the annulus, irregular, smooth, fleshy, hollow, continuous with the substance of the pileus. The **annulus** is formed of a few threads, remnants of the veil, which are stained black by the spores. Figure 29 is from plants (No. 3202 C. U. herbarium) collected near Ithaca, July 18, 1899.

[Pg 31]

Plate 8, Figure 29.—Hypholoma rugocephalum (7/8 natural size). Cap tawny, gills purple black, spotted. Copyright.

STROPHARIA Fr.

The genus *Stropharia* has purple-brown spores, the gills are attached to the stem, and the veil forms a ring on the stem.

Figure 30.—Stropharia semiglobata (natural size). Cap and stem light yellow, viscid, gills brownish purple. Copyright.

Stropharia semiglobata Batsch.—This species is rather common and widely distributed, occurring in grassy places recently manured, or on dung. The plants are scattered or clustered, rarely two or three joined at the base. They are 5–12 cm. high, the cap 1–3 cm. broad, and the stems 2–4 mm. in thickness. The entire plant is light yellow, and viscid when moist, the gills becoming purplish brown, or nearly black. Stevenson says it is regarded as poisonous. [Pg 32]

The **pileus** is rounded, then hemispherical (semi-globate), smooth, fleshy at the center, thinner toward the margin, even, very viscid or viscous when moist, light yellow. The **gills** are squarely set against the stem (adnate), broad, smooth, in age purplish brown to blackish, the color more or less clouded. The **spores** in mass, are brownish purple. The **stem** is slender, cylindrical, becoming hollow, straight, even or bulbous below, yellowish, but paler at the apex where there are often parallel striæ, marks from the gills in the young stage. The stem is often viscid and smeared with the glutinous substance which envelopes the plant when young, and from the more or less glutinous veil. The **ring** is glutinous when moist.

Figure 30 is from plants (No. 4613 C. U. herbarium) collected on one of the streets of Ithaca.

Stropharia stercoraria Fr., is a closely related plant, about the same size, but the pileus, first hemispherical, then becoming expanded and sometimes striate on the margin, while the stem is stuffed. The gills are said to be of one color and the ring floccose, viscose, and evanescent in drying. It occurs on dung, or in grassy places recently manured.

Stropharia æruginosa Curt., the greenish *Stropharia*, is from 6–8 cm. high, and the pileus 5–7 cm. broad. The ground color is yellowish, but the plant is covered with a greenish slime which tends to disappear with age. It is found in woods and open places during late summer and in autumn. According to Stevenson it is poisonous.

FOOTNOTES:

[B] For analytical key to the genera see Chapter XXIV .

CHAPTER V.

THE BLACK-SPORED AGARICS.

The spores are black in mass, not purple tinged. For analytical keys to the genera see Chapter XXIV .

COPRINUS Pers.

The species of *Coprinus* are readily recognised from the black spores in addition to the fact that the gills, at maturity, dissolve into a black or inky fluid. The larger species especially form in this way an abundance of the black fluid, so that it drops from the pileus and blackens the grass, etc., underneath the plant. In some of the [Pg 33] smaller species the gills do not wholly deliquesce, but the cap splits on top along the line of the longer gills, this split passing down through the gill, dividing it into two thin laminæ, which, however, remain united at the lower edge. This gives a fluted appearance to the margin of the pileus, which is very thin and membranaceous.

Figure 31. — Coprinus comatus, "shaggy-mane," in lawn.

The plants vary in size, from tiny ones to those which are several inches high and more than an inch broad. Their habitat (that is, the place where they grow) is peculiar. A number of the species grow on dung or recently manured ground. From this peculiarity the genus received the name *Coprinus* from the Greek word κοπρός, meaning dung. Some of the species, however, grow on decaying logs, on the ground, on leaves, etc.

Coprinus comatus Fr. **Edible.**—One of the finest species in this genus is the shaggy-mane, or horse-tail mushroom, as it is popularly called. It occurs in lawns and other grassy places, especially in richly manured ground. The plants sometimes occur singly, or a few together, but often quite large numbers of them appear in a small area. They occur most abundantly during quite wet weather, [Pg 34] or after heavy rains, in late spring or during the autumn, and also in the summer. From the rapid growth of many of the mushrooms we are apt to be taken by surprise to see them all up some day, when the day before there were none. The shaggy-mane often furnishes a surprise of this kind. In our lawns we are accustomed to a pretty bit of greensward with clumps of shrubbery, and here and there the overhanging branches of some shade tree. On some fine morning when we find a whole flock of these shaggy-manes, which have sprung up during the night, we can imagine that some such kind of a surprise must have come to Browning when he wrote these words:

"By the rose flesh mushroom undivulged

Last evening. Nay, in to-day's first dew

Yon sudden coral nipple bulged,

Where a freaked, fawn colored, flaky crew

Of toadstools peep indulged."

Figure 32.—Coprinus comatus. "Buttons," some in section showing gill slits and hollow stem; colors white and black. (Natural size.)

The plant is called shaggy-mane because of the very shaggy appearance of the cap, due to the surface being torn up into long locks. The illustrations of the shaggy mane shown here represent the different stages of development, and the account here given is largely taken from the account written by me in Bulletin 168 of the Cornell University Agr. Exp. Station.

Figure 33.—Coprinus comatus (natural size).]

[Pg 35] In Fig. 32 are shown two buttons of the size when they are just ready to break through the soil. They appear mottled with dark and white, for the outer layer of fungus threads, which are dark brown, is torn and separated into patches or scales, showing between the delicate meshes of white threads which lie beneath. The upper part of the button is already forming the cap, and the slight constriction about midway shows the lower boundary or margin of the pileus where it is still connected with the undeveloped stem.

At the right of each of these buttons in the figure is shown a section of a plant of the same age. Here the parts of the plant, though [Pg 36] still undeveloped, are quite well marked out. Just underneath the pileus layer are the gills. In the section one gill is exposed to view on either side. In the section of the larger button the free edge of the gill is still closely applied to the stem, while in the small one the gills are separated a short distance from the stems showing "gill slits." Here, too, the connection of the margin of the pileus with the stem is still shown, and forms the veil. This kind of a veil is a marginal veil.

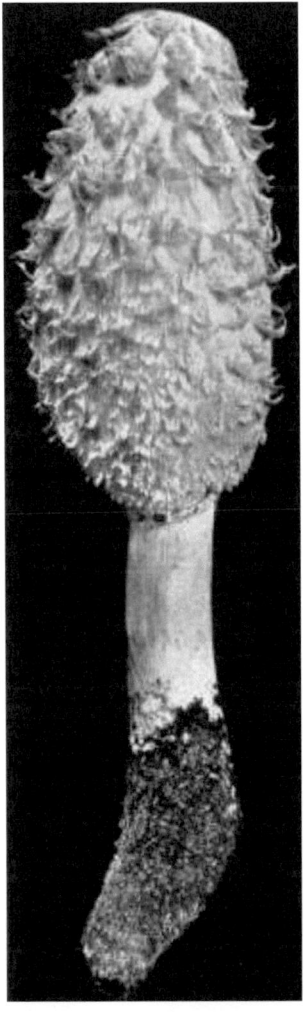

Figure 34.—Coprinus comatus (natural size). This one entirely white, none of the scales black tipped.

The stem is hollow even at this young stage, and a slender cord of mycelium extends down the center of the tube thus formed, as is shown in the sections.

The plants are nearly all white when full grown. The brown scales, so close together on the buttons, are widely separated except at the top or center of the pileus, where they remain close together and form a broad cap.

A study of the different stages, which appear from the button stage to the mature plant, reveals the cause of this change in color and the wide separation of the dark brown scales. The threads of the outer layer of the pileus, and especially those in the brown patches seen on the buttons, soon cease to grow, though they are firmly entangled with the inner layers. Now the threads underneath and all through the plant, in the gills and in the upper part of the stem, grow and elongate rapidly. This pulls on the outer layer, tearing it in the first place into small patches, and causing them later to be more widely separated on the mature plant. Some of these scales remain quite large, while others are torn up into quite small tufts.

Figure 35.—Coprinus comatus, sections of the plants in Fig. 33 (natural size).

[Pg 37] As the plant ages, the next inner layers of the pileus grow less rapidly, so that the white layer beneath the brown is torn up into an intricate tangle of locks and tufts, or is frazzled into a delicate pile which exists here and there between well formed tufts. While all present the same general characters there is considerable individual variation, as one can see by comparing a number of different plants. Figure 34 shows one of the interesting conditions. There is little of the brown color, and the outer portion of the pileus is torn into long locks, quite evenly distributed and curled up at the ends in an interesting [Pg 38] fashion which merits well the term "shaggy." In others the threads are looped up quite regularly into triangular tresses which appear to be knotted at the ends where the tangle of brown threads holds them together.

Figure 36.—Coprinus comatus, early stages of deliquescence; the ring is lying on the sod (natural size).

There is one curious feature about the expansion of the pileus of the shaggy-mane which could not escape our attention. The pileus has become very long while comparatively little lateral expansion has taken place. The pileus has remained cylindrical or barrel-shaped, while in the case of the common mushroom the pileus expands into the form of an umbrella.

Figure 37.—Coprinus comatus, later stage of deliquescence, pileus becoming more expanded (natural size).

The cylindrical or barrel-shaped pileus is characteristic of the shaggy-mane mushroom. As the pileus elongates the stem does also, but more rapidly. This tears apart the connection of the margin of the pileus with the base of the stem, as is plainly shown in Fig. 33. In breaking away, the connecting portion or veil is freed both from the stem and from the margin of the pileus, and is left as a free, or loose, ring around [Pg 39] the stem. In the shaggy-mane the veil does not form a thin, expanded curtain. It is really an annular outer layer of the button lying between the margin of the cap and the base of the stem. It becomes free from the stem. As the stem elongates more rapidly than the cap, the latter is lifted up away from the base of the stem. Sometimes the free ring is left as a collar around the base of the stem, still loosely adherent to the superficial layer of the same, or it remains for a time more or less adherent to the margin of the pileus as shown in the plant at the left hand in Fig. 33. It is often lifted higher up on the stem before it becomes free from the cap, and is then left dangling somewhere on the stem, or it may break and fall down on the sod. In other instances it may remain quite firmly adherent to the margin of the pileus so that it breaks apart as the pileus in age expands somewhat. In such cases one often searches for some time to discover it clinging as a sterile margin of the cap. It is interesting to observe a section of the plants at this stage. These sections can be made by splitting the pileus and stem lengthwise through the middle line with a sharp knife, as shown in Fig. 35. Here, in the plant at the right hand, the "cord" of mycelium is plainly seen running through the hollow stem. The gills form a large portion of the plant, for they are very broad and lie closely packed side by side. They are nowhere attached to the stem, but at the upper end round off to the cap, leaving a well defined space between their ends and the stem. The cap, while it is rather thick at the center, i. e., where it joins the stem, becomes comparatively thin where it spreads out over the gills. At this age of the plant [Pg 40] the gills are of a rich salmon color, i. e., before the spores are ripe, and the taste when raw is a pleasant nutty flavor, reminding one of the meat of fresh green hickory nuts. In a somewhat earlier stage the edges of all the gills are closely applied to the stem which they surround. So closely are they applied to the stem in most cases that threads of mycelium pass from the stem to the edge of the gills. As the cap expands slightly in ageing, these threads are torn asunder and the stem is covered with a very delicate down or with flocculent particles which easily disappear on handling or by the washing of the rains. The edges of the gills are also left in a frazzled condition, as one can see by examining them with a good hand lens.

The spores now begin to ripen and as they become black the color of the gills changes. At the same time the gills and the cap begin to dissolve into an inky fluid, first becoming dark and then melting into a black liquid. As this accumulates it forms into drops which dangle from the cap until they fall away. This change takes place on the margin of the cap first, and advances toward the center, and the contrast of color, as the blackening invades the rich salmon, is very striking. The cap now begins to expand outward more, so that it becomes somewhat umbrella shaped. The extreme outer surface does not dissolve so freely, and the thin remnant curls upward and becomes enrolled on the upper side as the cap with wasted gills becomes nearly flat.

Coprinus atramentarius (Bull.) Fr. **Edible.** — The ink-cap (*Coprinus atramentarius*) occurs under much the same conditions as the shaggy-mane, and is sometimes found accompanying it. It is usually more common and more abundant. It springs up in old or newly made lawns which have been richly manured, or it occurs in other grassy places. Sometimes the plants are scattered, sometimes two or three in a cluster, but usually large clusters are formed where ten to twenty or more are crowded closely together (Fig. 39). The stems are shorter than those of the shaggy-mane and the cap is different in shape and color. The cap is egg-shaped or oval. It varies in color from a silvery grey, in some forms, to a dark ashen grey, or smoky brown color in others. Sometimes the cap is entirely smooth, as I have seen it in some of the silvery grey forms, where the delicate fibres coursing down in lines on the outer surface cast a beautiful silvery sheen in the light. Other forms present numerous small scales on the top or center of the cap which are formed by the cleavage of the outer surface here into large numbers of pointed tufts. In others, the delicate tufts cover more or less the entire surface, giving the plant a coarsely granular aspect. This is perhaps the more common appearance, at least so far as my observation goes. But not infrequently one finds forms which have the entire outer surface of the cap torn into quite a large number of coarse scales, and these are often more prominent over the upper portion. Fine lines or striations mark also the entire surface of all the forms, especially toward the margin, where the scales are not so prominent. The marginal half of the cap is also frequently furrowed more or less irregularly, and this forms a crenate or uneven edge.

Plate 9, Figure 38. Coprinus comatus, drops of inky fluid about to fall from wasted pileus (natural size).

[Pg 41]

Plate 10, Figure 39.—Coprinus atramentarius, nearly smooth form, gray color (natural size)

Figure 40. — Coprinus atramentarius, scaly form (natural size).

The annulus or ring on the stem of the ink-cap is very different from that of the shaggy-mane. It forms an irregularly zigzag elevated line of threads which extend around the stem near the base. It is well shown in Fig. 41 as a border line between the lower scaly end of the stem and the smooth white upper part. It is formed at the time of the separation of the margin of the cap from the stem, the connecting fibres being pulled outward and left to mark the line [Pg 42] of junction, while others below give the scaly appearance. It is easily effaced by rough handling or by the washing of the rains. A section of a plant is illustrated by a photograph in Fig. 42. On either side of the stem is shown the layer of fibres which form the annulus, and this layer is of a different texture from that of the stem. The stem is hollow as seen here also. In this figure one can see the change in color of the gills just at the time when they begin to deliquesce. This deliquescence proceeds much in the same way as in the shaggy-mane, and sometimes the thin remnant of the cap expands and the margin is enrolled over the top.

Figure 41.—Coprinus atramentarius, showing annulus as border line between scaly and smooth part of the stem (natural size).

Coprinus micaceus (Bull.) Fr. **Edible.**—The glistening coprinus received its name because of the very delicate scales which often cover the surface of the cap, and glisten in the light like particles of mica. This plant is very common during the spring and early summer, though it does appear during the autumn. It occurs about the bases of stumps or trees or in grassy or denuded places, from dead [Pg 43] roots, etc., buried in the soil. It occurs in dense tufts of ten to thirty or more individuals; sometimes as many as several hundred spring up from the roots of a dead tree or stump along the streets or in lawns, forming large masses. More rarely it occurs on logs in the woods, and sometimes the plants are scattered in lawns. From the different habits of the plant it is sometimes difficult to determine, especially where the individuals are more or less scattered. However, the color, and the markings on the cap, especially the presence of the small shining scales when not effaced, characterize the plant so that little difficulty is experienced in determining it when one has once carefully noted these peculiarities.

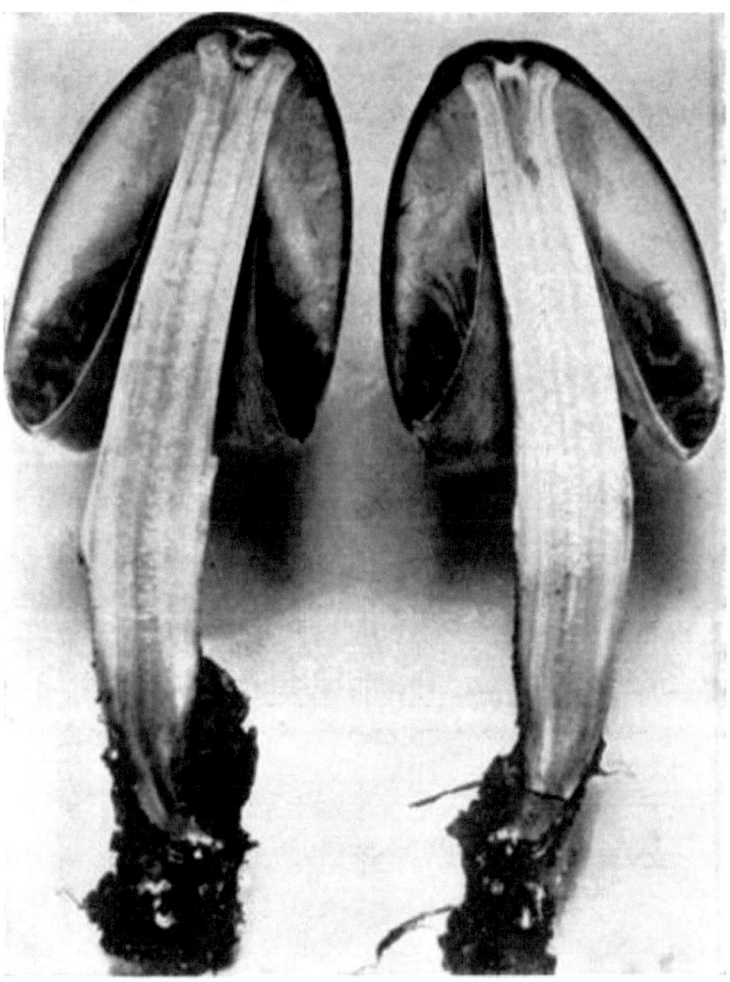

Figure 42.—Coprinus atramentarius, section of one of the plants in Fig. 41 (natural size).

Figure 43 is from a group of three young individuals photographed just as the margin of the pileus is breaking away from the lower part of the stem, showing the delicate fibrous ring which is formed in the same way as in *Coprinus atramentarius*. The ring is much more delicate and is rarely seen except in very young specimens which are carefully collected and which have not been washed by rains. The mature plants are 8–10 cm. high (3–4 inches), and the cap varies from 2–4 cm. in diameter. The stem is quite slender and the cap and gills quite thin as compared with the shaggy-mane and ink-cap. The gills are not nearly so crowded as they are in the two other [Pg 44] species. The cap is tan color, or light buff, or yellowish brown. Except near the center it is marked with quite prominent striations which radiate to the margin. These striations are minute furrows or depressed lines, and form one of the characters of the species, being much more prominent than on the cap of the ink-cap.

Figure 43.—Coprinus micaceus, young stage showing annulus, on the cap the "mica" particles (natural size).

Figure 44.—Coprinus micaceus, plants natural size, from floor of coal mine at Wilkesbarre. Caps tan color. Copyright.

In wet weather this coprinus melts down into an inky fluid also, but in quite dry weather it remains more or less firm, and sometimes it does not deliquesce at all, but dries with all parts well preserved, though much shrunken of course, as is the case with all the very fleshy fungi.

Plate 11, Figure 45.—Panæolus retirugis, group of plants from lawn along street, showing veil in young plants at the left, which breaks into V-shaped loops and clings to margin of the cap. Cap dark smoky color at first, becoming grayish in age (natural size). Copyright.

PANAEOLUS Fr. [Pg 45]

In *Panæolus*, the pileus is somewhat fleshy, or thin, the margin even, that is, not striate. The margin extends beyond the gills, and the gills are not uniform in color, being clouded or spotted with black and brown colors, the edge of the gills often white in contrast. The spores are black. The stem is usually smooth, sometimes floccose scaly, often long, firm, generally hollow. The veil is of interwoven threads, sometimes quite compact, especially when the plants are young. Peck, 23rd Report N. Y. State Mus., p. 10 et seq., gives a synopsis of five species.

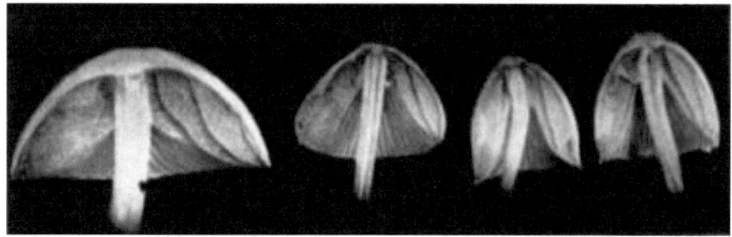

Figure 46. — Panæolus retirugis, section of caps showing form and position of gills (natural size).

Panæolus retirugis Fr. — The color of this plant is not attractive, but it is one of the most beautiful species I have studied, if one regards form and the general features in its development. It is said to occur on dung. I have found it in lawns or grassy places, especially freshly made lawns or greenswards which have been heavily manured. The illustrations in Figs. 45–48 were made from photographs of plants which grew in a newly made boulevard along Buffalo street, Ithaca, N. Y. (No. 2356 C. U. herbarium). The plants are from 7–15 cm. high, the cap from 1–3 cm. in diameter, and the stem is 3–4 mm. in thickness. The size of the plants varies greatly according to the environment, being larger in moist soil and wet weather and smaller in dry soil and dry weather. It occurs in late spring and during the summer.

Figure 47.—Panæolus retirugis, showing rugose character of cap in left-hand plant (natural size). Copyright.

The **pileus** is oval to ovate and conic, and in some cases it becomes more or less expanded, but never, so far as I have observed, does it become depressed or even plane. In wet weather it is usually at first dark smoky in color, viscid, becoming grayish in age, and as the pileus dries it becomes shining. In lighter colored forms the pileus is at first light leather color to cream color. Toward the center of the pileus are irregular wrinkles or shallow pits, the wrinkles anastomosing more or less, and it is because of this character of the surface of the pileus that the plant receives its specific name. During

79

dry weather there is a tendency for the pileus to crack, separating the dark color of the surface into patches showing the white flesh beneath. The pileus is often umbonate or gibbous, and the center is often darker than the margin. The pileus in rare cases is entirely white. The **gills** are adnate, broad in the middle, and in the more expanded forms as the gills separate more and more from the stem there is a tendency for them to become somewhat triangular. The **spores** are black in mass, are elliptical or short fusiform, and measure from 10–12 × 15–18 µ. The **stem** is cylindrical, [Pg 46] sometimes tortuous, smoky gray, light reddish brown, or paler, sometimes entirely white, the lighter forms of the stem accompanying the light forms of the pileus; cartilaginous in texture, becoming hollow, always darker below and paler above, smooth, granulate with minute darker points, bulbous. The **veil** is very [Pg 47] prominent and stout when the plant is young, and extends from the margin of the pileus to the stem when the plant is very young and the stem has not elongated. As the stipe elongates the veil separates from the stipe as a ring, and then, as the pileus expands, it is broken quite regularly into short segments which become arranged regularly around the margin of the pileus in the form of the letter V, which gives a beautiful appearance to this stage of the plant. It is only when the plants are fresh and moist that this condition of the veil can be seen, for on drying the veil collapses. Water is sometimes caught under the veil before the pileus separates far from the stem, and the spores falling thus float against the stem at this point and make a dark ring around the stem, which, however, should not be mistaken for the annulus. In no case was the veil observed to cling to the stem, and many plants have been observed to see if this variation might present itself.

Figure 48. — Panæolus retirugis, showing cracked surface of cap in the left-hand plant, also in same plant the ring mark of black spores which lodged before veil ruptured, in other plants showing well the V-shaped loops of veil on margin of cap (natural size). Copyright.

This peculiarity of the veil in clinging to the margin of the pileus [Pg 48] has led Hennings to place the plant in Karsten's genus (Engler and Prantl, Pflanzenfamilien) *Chalymotta*, as *Chalymotta retirugis*. The plants have several times been eaten raw by me, and while they have a nutty flavor and odor, the taste is not entirely agreeable in this condition, because of the accompanying slimy sensation.

A number of smaller species, among them **P. fimicola** Fr., and **P. papilionaceus** Fr., occur in similar places. **Panæolus solidipes** Pk., is a large species with a long, solid stem, growing on dung. **Psilocybe fœnisecii**, abundant in lawns and grassy places during late spring and summer, resembles a Panæolus. The cap shows zones of light and dark color, due to different amounts of water, which disappear as the plant matures. It belongs to the purple-brown-spored agarics.

PSATHYRELLA Fr.

The pileus is thin, membranaceous, striate, the margin not extending beyond the edge of the gills, and when young the margin of the pileus lies straight against the stem. The gills are black to fuliginous, of a uniform color, i. e., not spotted as in *Panæolus* and *Anellaria*. The spores are black. The plants are all fragile. Only one species is mentioned here. In appearance the species are like *Psathyra* of the purple-brown-spored agarics, but much thinner. Peck describes three species in the 23d Report N. Y. State Mus., p. 102 et seq. Only one species is described here.

Figure 49. — Psathyrella disseminata (natural size), caps whitish, grayish, or grayish-brown. Copyright.

Psathyrella disseminata Pers. — This is a very common and widely distributed species, appearing from late spring until late autumn. It [Pg 49] sometimes appears in greenhouses throughout the year. The plants are 2–3 cm. high, and the caps 6–10 mm. broad. The plants are crowded in large tufts, often growing on decaying wood, but also on the ground, especially about much decayed stumps, but also in lawns and similar places, where buried roots, etc., are decaying. They resemble small specimens of a *Coprinus*.

The **pileus** is whitish or gray, or grayish brown, very thin, oval, then bell-shaped, minutely scaly, becoming smooth, prominently silicate or plicate, plaited. The **gills** are adnate, broad, white, gray, then black. The **spores** are black, oblong, 8 × 6 μ. The **stem** is very slender, becoming hollow, often curved. The entire plant is very fragile, and in age becomes so soft as to suggest a *Coprinus* in addition to the general appearance. Figure 49 is from plants collected on decaying logs at Ithaca.

GOMPHIDIUS Fr.

The genus *Gomphidius* has a slimy or glutinous universal veil enveloping the entire plant when young, and for a time is stretched over the gills as the pileus is expanding. The gills are somewhat mucilaginous in consistency, are distant and decurrent on the stem. The gills are easily removed from the under surface of the pileus in some species by peeling off in strips, showing the imprint of the gills beneath the projecting portions of the pileus, which extended part way between the laminæ of the gills. The spores in some species are blackish, and for this reason the genus has been placed by many with the black-spored agarics, while its true relationship is probably with the genus *Hygrophorus* or *Paxillus*.

Gomphidius nigricans Pk. — The description given by Peck for this plant in the 48th Report, p. 12, 1895, reads as follows:

"Pileus convex, or nearly plane, pale, brownish red, covered with a tough gluten, which becomes black in drying, flesh firm, whitish; lamellæ distant, decurrent, some of them forked, white, becoming smoky brown, black in the dried plant; stem subequal, longer than the diameter of the pileus, glutinous, solid, at first whitish, especially at the top, soon blackish by the drying of the gluten, whitish within, slightly tinged with red toward the base; spores oblong fusoid, 15–25 μ long, 6–7 μ broad. Pileus 1–2 inches broad; stem 1.5–2.5 inches long, 2–4 lines thick."

"This species is easily known by the blackening gluten which smears both pileus and stem, and even forms a veil by which the lamellæ in the young plant are concealed. In the dried state the whole plant is black." [Pg 50]

"Under pine trees, Westport, September."

Figure 50.—Gomphidius nigricans. Side and under view showing forked gills, and reticulate collapsed patches of dark slime on stem. Cap flesh color, gills dark gray; entire plant black when dried (natural size). Copyright.

What appears to be the same plant was collected by me at Blowing Rock, N. C., under a pine tree, in September, 1899 (No. 3979 C. U. herbarium).

The notes taken on the fresh plant are as follows:

Very viscid, with a thick, tough viscid cuticle, cortina or veil viscid, and collapsing on the stem, forming coarse, walnut-brown or dark vinaceous reticulations, terminating abruptly near the gills, or reaching them.

The **stem** is white underneath the slimy veil covering, tough, fibrous, continuous, and not separable from the hymenophore, tapering below.

The **pileus** is convex, the very thin margin somewhat incurved, disk expanded, uneven, near the center cracked into numerous small viscid brownish areoles; pileus flesh color, flesh same color except toward the gills. Gills dark drab gray, arcuate, distant, decurrent, many of them forked, separating easily from the hymenophore, peeling off in broad sheets, and leaving behind corresponding elevations of the hymenophore [Pg 51] which extended between the laminæ of the lamellæ. Pileus 7 cm. in diameter; stem 4–5 cm. long by 2 cm. diameter.

In drying, the entire plant as well as the gluten becomes black, on the pileus a shining black.

The **spores** are rusty to dark brown, or nearly black, fusoid or oblong, and measure 15–22 × 5–6 µ.

Figure 51.—Gomphidius nigricans. Under view with portion of gills stripped off from hymenophore, showing forked character of gills (natural size). Copyright.

In Fig. 50 a side and under view of the plant are given, and in Fig. 51 a view after a portion of the lamellæ have been peeled off, showing how nicely the separation takes place, as well as showing the forked character of the lamellæ and the processes of the pileus, which extend between the laminæ of the lamellæ.

This plant seems to be very near *Gomphidius glutinosus* (Schaeff.), Fr., if not identical with it, though the illustrations cited in Schaeffer and in Krombholz seem to indicate a stouter plant. The descriptions say nothing as to the appearance of the dried plant.

CHAPTER VI. [Pg 52]

THE WHITE-SPORED AGARICS.

The spores are white in mass, or sometimes with a faint yellowish or lilac tinge. For analytical keys to the genera see Chapter XXIV .

AMANITA Pers.

The genus *Amanita* has both a volva and a veil; the spores are white, and the stem is easily separable from the cap. In the young stage the volva forms a universal veil, that is, a layer of fungus tissue which entirely envelops the young plant. In the button stage, where this envelope runs over the cap, it is more or less free from it, that is, it is not "concrete" with the surface of the pileus. As the pileus expands and the stem elongates, the volva is ruptured in different ways according to the species. In some the volva splits at the apex and is left as a "cup" at the base of the stem. In others it splits circularly, that is, transversely across the middle, the lower half forming a shallow cup with a very narrow rim, or in other cases it is closely fitted against the stem, while the upper half remains on the cap and is broken up into patches or warts. In still other cases the volva breaks irregularly, and only remnants of it may be found on either the base of the stem or on the pileus. For the various conditions one must consult the descriptions of the species. The genus is closely related to *Lepiota*, from which it is separated by the volva being separate from the pileus. This genus contains some of the most deadly poisonous mushrooms, and also some of the species are edible. Morgan, Jour. Mycol. **3**: 25–33, describes 28 species. Peck, 33d Report N. Y. State Mus., pp. 38–49, describes 14 species. Lloyd, A Compilation of the Volvæ of the U. S., Cincinnati, 1898, gives a brief synopsis of our species.

Amanita muscaria Linn. **Poisonous.** — This plant in some places is popularly known as the fly agaric, since infusions of it are used as a fly poison. It occurs during the summer and early autumn. It grows along roadsides near trees, or in groves, and in woods, according to some preferring a rather poor gravelly soil. It attains its typical form usually under these conditions in groves or rather open woods where the soil is poor. It is a handsome and striking plant because of the usually brilliant coloring of the cap in contrast with the white stems and gills, and the usually white scales on the surface. It usually ranges from 10–15 cm. high, and the cap from 8–12 cm. broad, while the stem is 1–1.5 cm. in thickness, or the plant may be considerably larger.

Plate 12, Figure 52. — Amanita muscaria, "buttons," showing different stages of rupture of the volva or universal veil, and formation of inner veil (natural size). Copyright.

[Pg 53]

Plate 13, Figure 53.—Amanita muscaria. Further stages in open-ing of plant, formation of veil and ring. Cap yellowish, or orange. Scales on cap and at base of stem white; stem and gills white (natu-ral size). Copyright.

The **pileus** passes from convex to expanded and nearly flat in age, the margin when mature is marked by depressed lines forming parallel striations, and on the surface are numbers of scattered floc-cose or rather compact scales, formed from the fragments of the upper part of the volva or outer veil. These scales are usually white in color and are quite easily removed, so that old plants are some-times quite free from them. The scales are sometimes yellowish in color. The color of the pileus varies from yellow to orange, or even red, the yellow color being more common. Late in the season the color is paler, and in old plants also the color fades out, so that white forms are sometimes found. The flesh is white, sometimes yellowish underneath the cuticle. The **gills** in typical forms are white, in some forms accredited to this species they are yellowish. The **stem** is cylindrical, hollow, or stuffed when young, and en-larged below into a prominent bulb. It is white, covered with loose floccose scales, or more or less lacerate or torn, and the lower part of the stem and upper part of the bulb are marked usually by promi-nent concentric scales forming interrupted rings. These are formed by the splitting of the outer veil or volva, and form the remnants of the volva present on the base of the stem.

The main features in the development of the plant are shown in Figs. 52–54, where a series from the button stage to the mature plant is represented. In the youngest specimens the outline of the bulb and the young convex or nearly globose cap are only seen, and these are covered with the more or less floccose outer veil or volva. The fungus threads composing this layer cease to grow, and with the expansion of the cap and the elongation of the stem, the volva is torn into patches. The upper and lower surface of the inner veil is attached to the edge of the gills and to the outer surface of the stem by loose threads, which are torn asunder as the pileus expands. Floccose scales are thus left on the surface of the stem below the annulus, as in the left hand plant of Fig. 53. The veil remains attached longer to the gills and is first separated from the stem. Again, as in the right hand plant, it may first be separated from the gills when it is later ripped up from the stem.

The fly agaric is one of the well known poisonous species and is very widely distributed in this country, as well as in other parts of [Pg 54] the world. In well developed forms there should be no difficulty in distinguishing it from the common mushroom by even a novice. Nor should there be difficulty in distinguishing it from the royal agaric, or Cæsar's agaric (*Amanita cæsarea*), by one who has become reasonably familiar with the characters and appearance of the two. But small and depauperate specimens of the two species run so nearly together in form, color, and surface characters, that it becomes a matter of some difficulty for even an expert to distinguish them.

Figure 54.—Amanita muscaria. View of upper side of cap (natural size). Colors as in Fig. 53. Copyright.

Figures 52–54 are from plants (No. 2065 C. U. herbarium) collected in an open woods near Ithaca. For the poisonous property of the plant see Chapter XXII .

Amanita frostiana Pk. **Poisonous.**—According to Dr. Peck, who published the first description of this plant, it grows in company with *Amanita muscaria*, but seems to prefer more dense woods, especially mixed or hemlock woods, and occurs from June to October. The plant is 5–8 cm. high, the caps 2–5 cm. broad, and the stems 3–6 mm. in thickness.

The **pileus** is "convex to expanded, bright orange or yellow, warty, sometimes nearly or quite smooth, striate on the margin; [Pg 55] **lamellæ** white or tinged with yellow; **stem** white or yellowish, stuffed, bearing a slight, sometimes evanescent annulus, bulbous at the base, the bulb *slightly margined* by the volva; spores globose," 7.5–10 µ in diameter. He notes that it appears like a small form of *A. muscaria*, to which it was first referred as *var. minor*,—"The only characters for distinguishing it are its small size and its globose spores." It is near *A. muscaria var. puella* Pers.

I have several times found this plant in the Adirondack mountains, N. Y., and Ithaca, and also at Blowing Rock, N. C. The volva is often yellowish, so that the warts on the pileus are also yellow, and sometimes the only remnants of the volva on the base of the stem are yellow or orange particles. The annulus is also frequently yellow. In our plants, which seem to be typical, the spores are nearly globose, varying to oval, and with the minute point where the spore was attached to the sterigma at the smaller end, the spores usually being finely granular, 6–9 μ in diameter, and rarely varying towards short elliptical, showing a tendency to approach the shape of the spores of *A. muscaria*. The species as I have seen it is a very variable one, large forms being difficult to separate from *A. muscaria*, on the one hand, and others difficult to separate from the depauperate forms of *A. cæsarea*. In the latter, however, the striæ are coarser, though the yellow color may be present only on portions of the pileus. The spores of *A. cæsarea* are from globose to oval, ovate or short elliptical, the globose ones often agreeing in size with the spores of *A. frostiana*, but they usually contain a prominent oil drop or "nucleus," often nearly filling the spore. In some specimens of *A. frostiana* the spores are quite variable, being nearly globose, ovate to elliptical, approaching the spores of *A. muscaria*. These intermediate forms should not in themselves lead one to regard all these three species as representing variations in a single variable species. With observations in the field I should think it possible to separate them.

Amanita phalloides Fr. **Deadly Poisonous.** — The *Amanita phalloides* and its various forms, or closely related species, are the most dangerous of the poisonous mushrooms. For this reason the *A. phalloides* is known as the *deadly agaric*, or *deadly amanita*. The plant is very variable in color, the forms being pure white, or yellowish, green, or olive to umber. Variations also occur in the way in which the volva ruptures, as well as in the surface characters of the stem, and thus it is often a difficult matter to determine whether all these forms represent a single variable species or whether there are several species, and if so, what are the limits of these [Pg 56] species. Whether these are recognized as different forms of one species or as different species, they are all very poisonous. The plant usually occurs in woods or along the borders of woods. It does, however, sometimes occur in lawns. It varies from 6–20 cm. high, the cap from 3–10 cm. broad, and the stem 6–10 mm. in thickness.

Figure 55.—Amanita phalloides, white form, showing cap, stem, ring, and cup-like volva with a free, prominent limb (natural size).

The **pileus** is fleshy, viscid or slimy when moist, smooth, that is, not striate, orbicular to bell-shaped, convex and finally expanded, and in old specimens more or less depressed by the elevation of the margin. The cap is often free from any remnants of the volva, while in other cases portions of the volva or outer veil appear on the surface of the cap in rather broad patches, or it may be broken up into a number of smaller ones quite evenly distributed over the surface of the cap. The presence or absence of these scales on the cap depends entirely on the way in which the volva ruptures. When there is a clean rupture at the apex the pileus is free from scales, but if portions of the apex of the volva are torn away they are apt to remain on the cap.

Plate 14, Figure 56. — Amanita phalloides, brownish, umber, or olive-brown form (natural size). Caps brownish or whitish, and streaked with brown, scales white, gills and stem white, stem slowly turning dull brown where bruised. Copyright.

The white form is common in this country, and so is the olive or umber form. The yellow form is rarer. Sometimes there is only a tinge of yellow at the center of the white pileus, while in other cases a large part of the pileus may be yellow, a deeper shade usually on the center. The green form is probably more common in Europe than in this country. The olive form varies considerably also in the depth of the color, usually darker on the center and fading out to light olive or gray, or whitish, on the margin. In other cases the [Pg 57] entire pileus may be dark olive or umber color. The **gills** in all the forms are white, and free from the stem or only joined by a narrow line. The stem is stuffed when young, but in age is nearly or quite hollow. It is cylindrical, 6–20 cm. long × 6–12 mm. in thickness. In the larger specimens the bulb is quite prominent and abrupt, while in the smaller specimens it is not always proportionally so large. The **stem** is usually smooth and the color is white, except in the dark forms, when it is dingy or partakes more or less of the color of the pileus, though much lighter in shade. There is a tendency in these forms to a discoloration of the stem where handled or bruised, and this should caution one in comparing such forms with the edible *A. rubescens*.

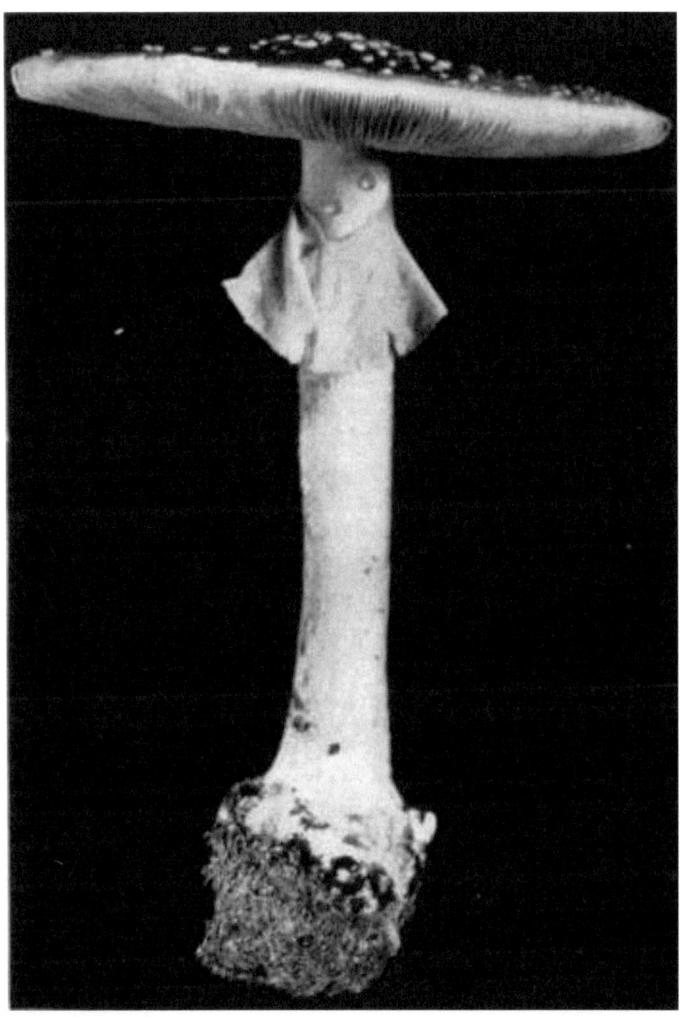

Figure 57.—Amanita phalloides, volva circumscissile, cap scaly, limb of volva not prominent, cap dark, scales white (natural size). Copyright.

Perhaps no part of the plant is more variable than the outer veil or volva. Where the volva is quite thick and stout it usually splits at the apex, and there is a prominent free limb, as shown in Fig. 55.

95

Sometimes thin portions of the volva are caught, and remain on the surface of the pileus. But when the volva is thinner and of a looser texture, it splits transversely about the middle, circumscissile, and all or a large part of the upper half of the volva then clings to the cap, and is separated into patches. Between this and the former condition there seem to be all gradations. Some of these are shown [Pg 58] in Fig. 56, which is from a photograph of dark olive and umber forms, from plants collected in the Blue Ridge mountains, at Blowing Rock, N. C., during September, 1899. In the very young plant the volva split transversely (in a circumscissile fashion) quite clearly, and the free limb is quite short and distant from the stem on the margin of the saucer-like bulb. In the large and fully expanded plant at the center, the volva ruptured irregularly at the apex, and portions of the thin upper half remain as patches on the cap while the larger part remains as the free limb, attached at the margin of the broad saucer-shaped bulb, and collapsed up against the base of the stem.

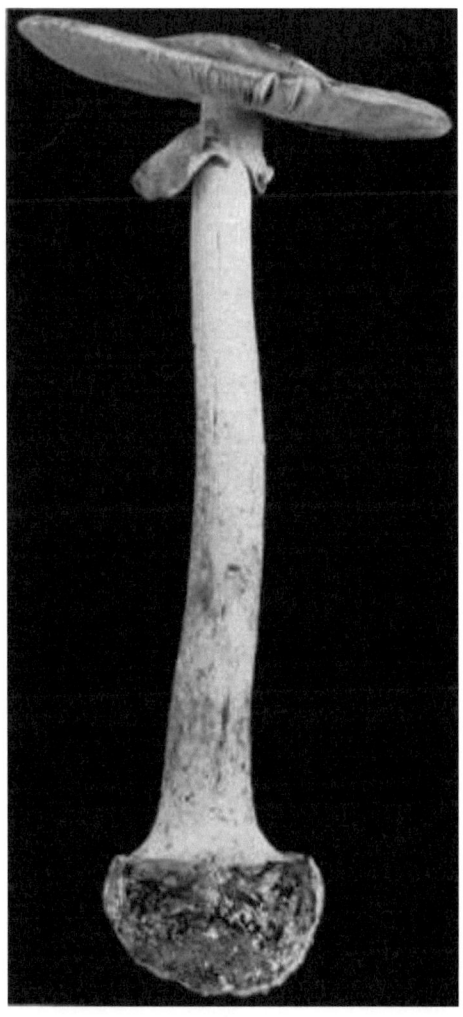

Figure 58.—Amanita phalloides, volva circumscissile, concave bulb margined by definite short limb of volva; upper part of volva has disappeared from cap; cap whitish, tinged with brown.

Figure 58 and the small plant in Fig. 56, both from photographs of the sooty form of *Amanita phalloides*, show in a striking manner the typical condition of the circumscissile volva margining the broad |

saucer-like bulb as described for *Amanita mappa*. The color of *A. mappa* is usually said to be straw color, but Fries even says that the color is as in *A. phalloides*, "now white, now green, now yellow, now dark brown" (Epicrisis, page 6). According to this, Fig. 58 would represent *A. mappa*.

The variable condition in this one species *A. phalloides*, now splitting at the apex, now tearing up irregularly, now splitting in a definitely circumscissile manner, seems to [Pg 59] bid defiance to any attempt to separate the species of *Amanita* into groups based on the manner in which the volva ruptures. While it seems to be quite fixed and characteristic in certain species, it is so extremely variable in others as to lead to the suspicion that it is responsible in some cases for the multiplication and confusion of species. At the same time, the occurrence of some of these forms at certain seasons of the year suggests the desirability of prolonged and careful study of fresh material, and the search for additional evidence of the unity of these forms, or of their definite segregation.

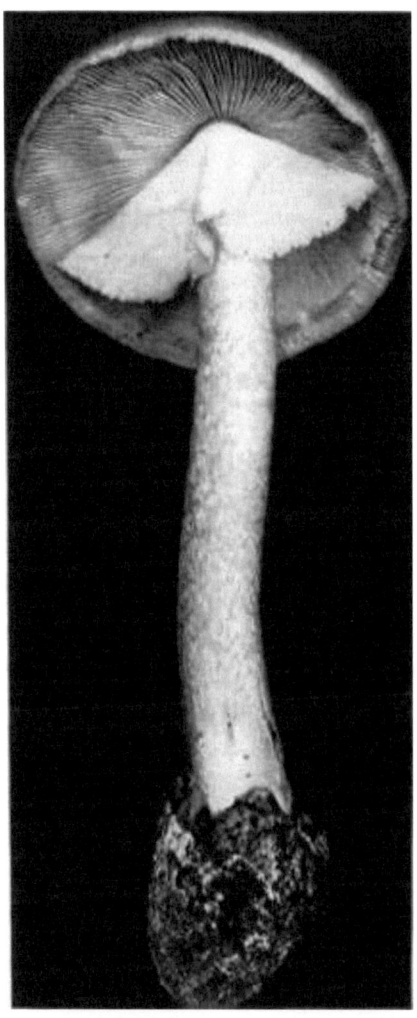

Figure 59. — Amanita verna, white (natural size). Copyright.

Since the *Amanita phalloides* occurs usually in woods, or along borders of woods, there is little danger of confounding it with edible mushrooms collected in lawns distant from the woods, and in open fields. However, it does occur in lawns bordering on woods, and in the summer of 1899 I found several of the white forms of this species in a lawn distant from the woods. This should cause beginners and those not thoroughly familiar with the appearance of the plant to be extremely cautious against eating mushrooms simply because they were not collected in or near the woods. Furthermore, sometimes the white form of the deadly amanita possesses a faint tinge of pink in the gills, which might lead the novice to mistake it for the common [Pg 60] mushroom. The bulb of the deadly amanita is usually inserted quite deep in the soil or leaf mold, and specimens are often picked leaving the very important character of the volva in the ground, and then the plant might easily be taken for the common mushroom, or more likely for the smooth lepiota, *Lepiota naucina*, which is entirely white, the gills only in age showing a faint pink tinge. It is very important, therefore, that, until one has such familiarity with these plants that they are easily recognized in the absence of some of these characters, the stem should be carefully dug from the soil. In the case of the specimens of the deadly amanita growing in the lawn on the campus of Cornell University, the stems were sunk to three to four inches in the quite hard ground.

Amanita verna Bull. **Deadly Poisonous.**—The *Amanita verna* is by some considered as only a white form of the *Amanita phalloides*. It is of a pure white color, and this in addition to its very poisonous property has led to its designation as the "destroying angel."

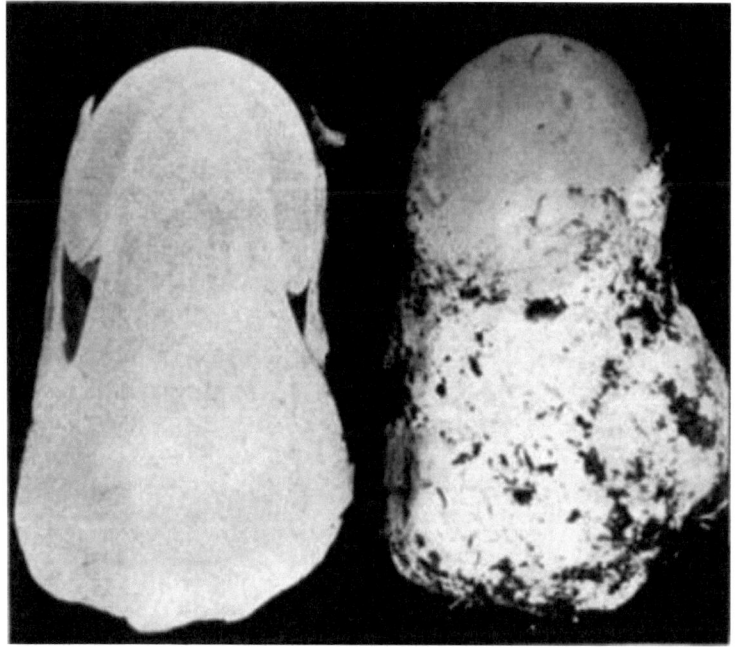

Figure 60.—Amanita verna, "buttons," cap bursting through the volva; left hand plant in section (natural size). Copyright.

The **pileus** is smooth and viscid when moist; the gills free; the **stem** stuffed or hollow in age; the **annulus** forms a broad collar, and the **volva** is split at the apex, and being quite stout, the free limb is prominent, and it hugs more or less closely to the base of the stem. Figure 59 represents the form of the plant which Gillet recognizes as *A. verna*; the pileus convex, the annulus broad and entire, and the stem scaly. These floccose scales are formed as a result of the separation of the annulus from the outer layer of the stem.

The characters presented in the formation of the veil and annulus in this species are very interesting, and sometimes present two of the types in the formation of the veil and annulus found in the genus *Amanita*. In the very young plant, in the button stage, as the young gills lie with their edges close against the side of the stem, [Pg 61] loose threads extend from the edges of the gills to the outer layer of the stem. This outer layer of the stem forms the veil, and is

101

more or less loosely connected with the firmer portion of the stem by loose threads. As the pileus expands, the threads connecting the edges of the gills with the veil are stronger than those which unite the veil with the surface of the stem. The veil is separated from the stem then, simultaneously, or nearly so, throughout its entire extent, and is not ripped up from below as in *Amanita velatipes*.

As the pileus expands, then, the veil lies closely over the edges of the gills until finally it is freed from them and from the margin of the pileus. As the veil is split off from the surface of the stem, the latter is torn into numerous floccose scales, as shown in Fig. 59.

In other cases, in addition to the primary veil which is separated from the stem in the manner described above, there is a secondary veil formed in exactly the same way as that described for *Amanita velatipes*.

Figure 61.—Amanita verna, small form, white (natural size). Copyright.

In such cases there are two veils, or a double veil, each attached to the margin of the pileus, the upper one ascending over the edges of the gills and attached above on the stem, while the lower one descends and is attached below as it is being ripped up from a second

103

layer of the stem. Figures 59–61 are from plants collected at Blowing Rock, N. C., in September, 1899.

Amanita virosa Fr. **Deadly Poisonous.**—This plant also by some is regarded as only a form of *Amanita phalloides*. It is a pure white plant and the pileus is viscid as in the *A. verna* and *A. phalloides*. The volva splits at the apex as in *A. verna*, but the veil is very fragile and torn into shreds as the pileus expands, portions of it clinging [Pg 62] to the margin of the cap as well as to the stem, as shown in Fig. 62. The stem is also adorned with soft floccose scales. Gillet further states that the pileus is conic to campanulate, not becoming convex as in *A. verna* and *A. phalloides*.

The variability presented in the character of the veil and in the shape of the pileus suggests, as some believe, that all these are but forms of a single variable species. On the other hand, we need a more careful and extended field study of these variations. Doubtless different interpretations of the specific limits by different students will lead some to recognize several species where others would recognize but one. Since species are not distinct creations there may be tolerably good grounds for both of these views.

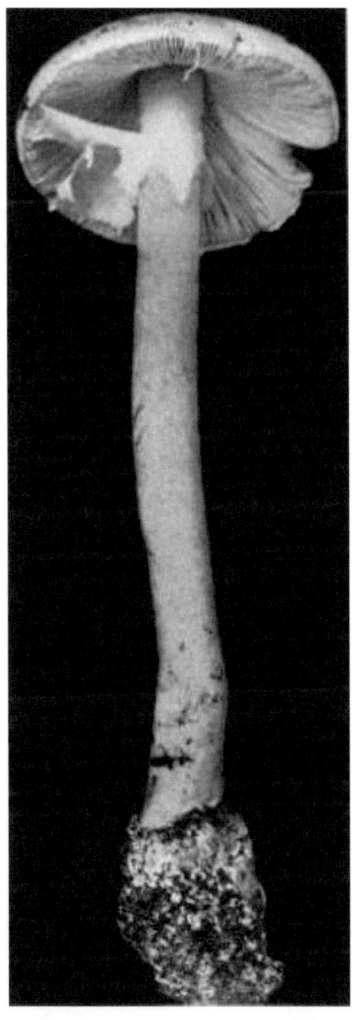

Figure 62. — Amanita virosa, white (natural size). Copyright.

Amanita floccocephala Atkinson. **Probably Poisonous.** — This species occurs in woods and groves at Ithaca during the autumn. The plants are medium sized, 6–8 cm. high, the cap 3–6 cm. broad, and the stems 4–6 mm. in thickness.

The **pileus** is hemispherical to convex, and expanded, smooth, whitish, with a tinge of straw color, and covered with torn, thin floccose patches of the upper half of the circumscissile volva. The **gills** are white and adnexed. The **spores** are globose, 7–10 μ. The **stem** is cylindrical or slightly tapering above, hollow or stuffed, [Pg 63] floccose scaly and abruptly bulbous below. The **annulus** is superior, that is, near the upper end of the stem, membranaceous, thin, sometimes tearing, as in *A. virosa*. The **volva** is circumscissile, the margin of the bulb not being clear cut and prominent, because there is much refuse matter and soil interwoven with the lower portion of the volva. The bulb closely resembles those in Cooke's figure (Illustrations, 4) of *A. mappa*. Figure 63 shows these characters well.

Figure 63. — Amanita floccocephala (natural size). Copyright.

Amanita velatipes Atkinson. **Properties Unknown.** — This plant is very interesting since it shows in a striking manner the peculiar way in which the veil is formed in some of the species of *Amanita*. Though not possessing brilliant colors, it is handsome in its form and in the peculiar setting of the volva fragments on the rich brown or faint yellow of the pileus. It has been found on several occasions during the month of July in a beech woods on one of the old flood plains of Six-mile creek, one of the gorges in the vicinity of Ithaca, N. Y. The mature plant is from 15–20 cm. high, the cap from 8–10 cm. broad, and the stem 1–1.5 cm. in thickness.

The **pileus** is viscid when moist, rounded, then broadly oval and convex to expanded, striate on the margin, sometimes in old [Pg 64] plants the margin is elevated. It is smooth throughout, and of a soft, rich hair brown, or umber brown color, darker in the center. Sometimes there is a decided but dull maize yellow tinge over the larger part of the pileus, but even then the center is often brown in color, shading into the yellow color toward the margin; the light yellow forms in age, often thinning out to a cream color. The flesh of the pileus is rather thin, even in the center, and becomes very thin toward the margin, as shown in Fig. 67. The scales on the pileus are more or less flattened, rather thin, clearly separated from the pileus, and easily removed. They are more or less angular, and while elongated transversely at first, become nearly isodiametric as the pileus becomes fully expanded, passing from an elongated form to rectangular, or sinuous in outline, the margin more or less upturned, especially in age, when they begin to loosen and "peel" from the surface of the cap. They are lighter in color than the pileus and I have never observed the yellow tint in them. The **gills** are white, broad at the middle, about 1 cm., and taper gradually toward each end. The **spores** are usually inequilaterally oval, 8–10 × 6–7 μ, granular when young, when mature with a large oil drop.

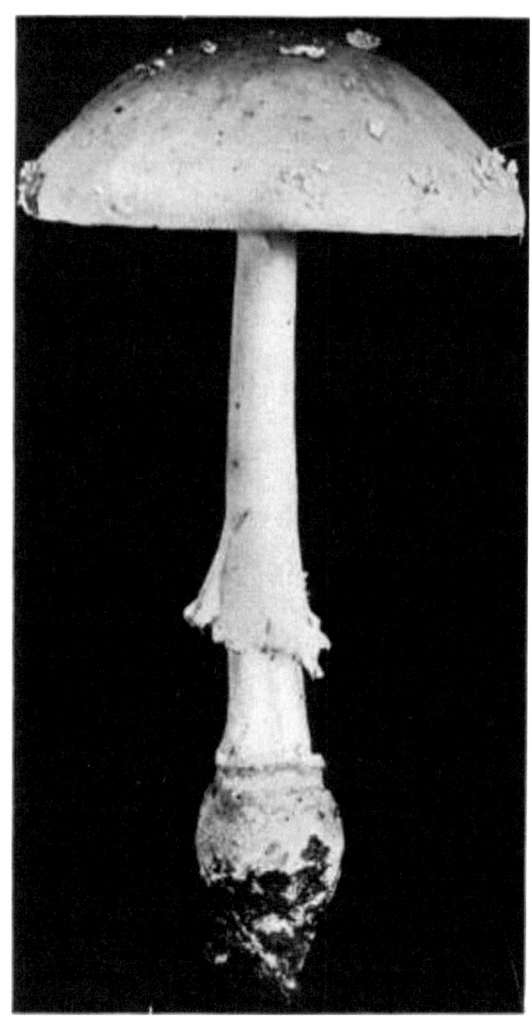

Plate 15, Figure 64.—Amanita velatipes (3/4 natural size). Cap hair-brown, or umber-brown, sometimes with tinge of lemon yellow, or entirely maize-yellow. Scales, gills, and stem white. Copyright.

Figure 65. — Amanita velatipes. Different stages of "buttons," in the right-hand plant the upper part of the volva separating to form the scales (natural size). Copyright.

The **stem** is cylindrical, somewhat bulbous, the bulb often tapering abruptly, as shown in Figs. 64, 66. The stem is white, smooth, or floccose scaly where the veil has been ripped off from it. It is [Pg 65] hollow and stuffed with loose cottony threads, as shown in Fig. 67. The **veil** is formed by the ripping up of the outer layer of the stem as the latter elongates and as the pileus expands. When it is freed from the margin of the cap it collapses and hangs downward as a broad collar (Fig. 64). The **annulus** is inferior, its position on the stem being due to the peculiar way in which it is formed.

Figure 66. — Amanita velatipes. Three plants natural size, the left-hand one sectioned, showing stuffed center of stem. Others show how veil is ripped up from the stem. For other details see text. Copyright.

Some of the stages of development are illustrated in Figs. 64–67. The buttons are queer looking objects, the bulb being the most prominent part. It tapers abruptly below, and on the upper side is the small rounded young cap seated in the center. The volva is present as a rough floccose layer, covering the upper part of the bulb [Pg 66] and the young cap. As the stem elongates and the pileus enlarges and expands, the volva is torn into areolate patches. The lower patches, those adjoining the margin of the cap and the upper part of the bulb, are separated in a more or less concentric manner. One or more of them lie on the upper part of the bulb, forming the

110

"limb" of the "ocreate" volva. Others lie around the margin of the pileus. Sometimes an annular one bordering the pileus and bulb is left clinging part way up on the stem, as shown in Fig. 66. The concentric arrangement on the pileus is sometimes shown for a considerable time, as in Fig. 67, the elongated areas being present in greater number at this age of the pileus. However, as the pileus expands more, these are separated into smaller areas and their connection with the surface of the pileus becomes less firm.

The formation of the veil and annulus can be easily followed in these figures. The margin of the cap in the button stage is firmly connected with the outer layer of the stem at its lower end. This probably occurs by the intermingling growth of the threads from the lower end of the stem and the margin of the cap, while the edges of the gills are quite free from the stem. Now as the stem elongates and the cap expands the veil is "ripped" up from the outer part of the stem. This is very clearly shown in Fig. 66, especially where two strips on the stem have become disconnected from the margin of the cap and are therefore left in position on the outside of the stem.

This species is related to *A. excelsa* Fr., which is said to have a superior ring.

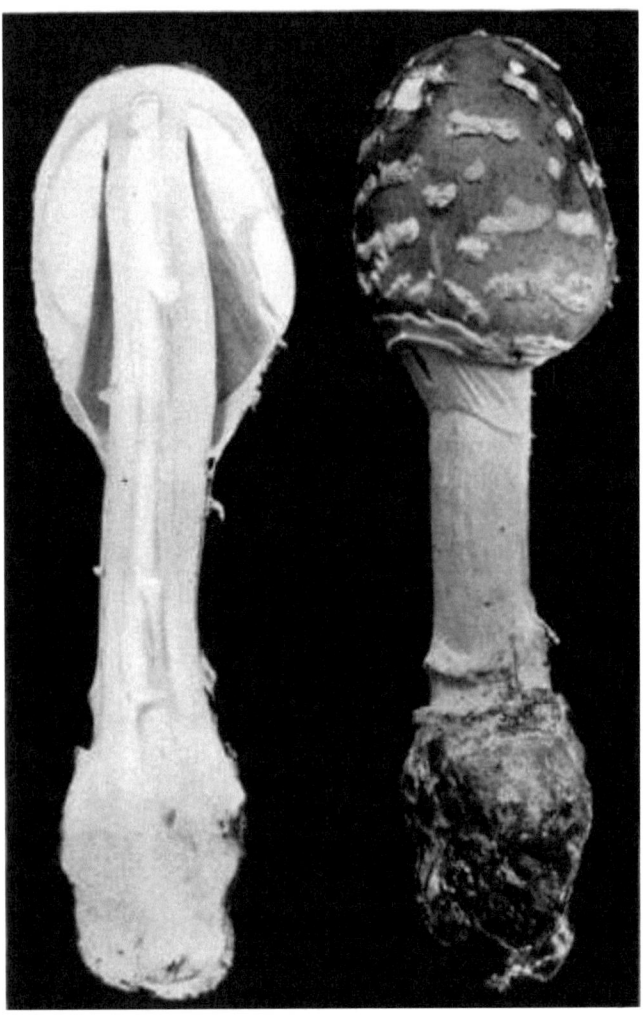

Plate 16, Figure 67.—Amanita velatipes. The right-hand plant shows how the veil is ripped up from the stem and also shows the transversely elongate scales on the cap. For details see text (natural size). Copyright.

Amanita cothurnata Atkinson. **Probably Poisonous.**—The booted amanita, *Amanita cothurnata*, I have found in two different years in the Blue Ridge mountains at Blowing Rock, N. C., once in 1888, during the first week of September, and again during the three first weeks in September, 1899. It occurs sparingly during the first week or so of September, and during the middle of the month is very abundant. The species seems to be clearly distinct from other species of *Amanita*, and there are certain characters so persistent as to make it easily recognizable. It ranges in height from 7–12 cm. and the caps are 3–7 cm. or more broad, while the stems are 4–10 mm. in thickness. The entire plant is usually white, but in some specimens the cap has a tinge of citron yellow, or in others tawny olive, in the center.

Plate 17, Figure 68.—Amanita cothurnata. Different stages of development; for details see text. Entire plant white, sometimes tinge of umber at center of cap, and rarely slight tinge of lemon-yellow at center (natural size). Copyright.

The **pileus** is fleshy, and passes, in its development, from nearly globose to hemispherical, convex, expanded, and when specimens are very old sometimes the margin is elevated. It is usually white, though specimens are found with a tinge of citron yellow in the [Pg 67] center, or of tawny olive in the center of other specimens. The pileus is viscid, strongly so when moist. It is finely striate on the margin, and covered with numerous, white, floccose scales from the upper half of the volva, forming more or less dense patches, which may wash off in heavy rains. The **gills** are rounded next the stem, and quite remote from it. The edge of the gills is often eroded or frazzly from the torn out threads with which they were loosely connected to the upper side of the veil in the young or button stage. The **spores** are globose or nearly so, with a large "nucleus" nearly filling the spore.

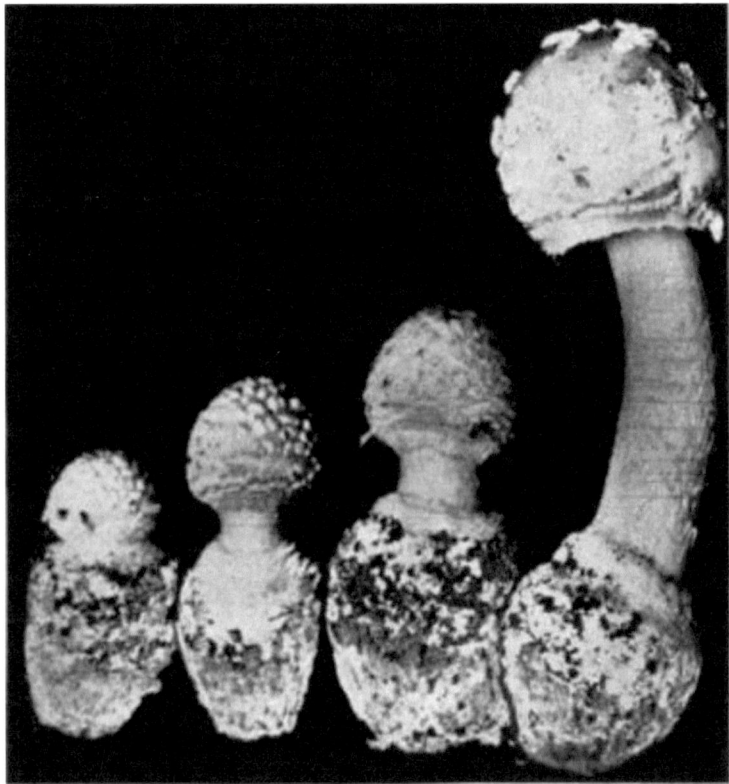

Figure 69.—Amanita cothurnata. Different stages opening up of plant, the two center ones showing veil being ripped from stem, but veil narrow. The right-hand illustration has been scratched transversely, these marks not being characteristic of the plant (natural size). Copyright.

The **stem** is cylindrical, even, and expanded below into quite a large oval bulb, the stem just above the bulb being margined by a close fitting roll of the volva, and the upper edge of this presenting the appearance of having been sewed at the top like the rolled edge of a garment or buskin. The surface of the stem is minutely floccose scaly or strongly so, and decidedly hollow even from a very young stage, or sometimes when young with loose threads in the cavity.

Figures 68–70, from plants (No. 3715, C. U. herbarium) collected at Blowing Rock, N. C., during September, 1899, illustrate certain of the features in the form and development of this plant.

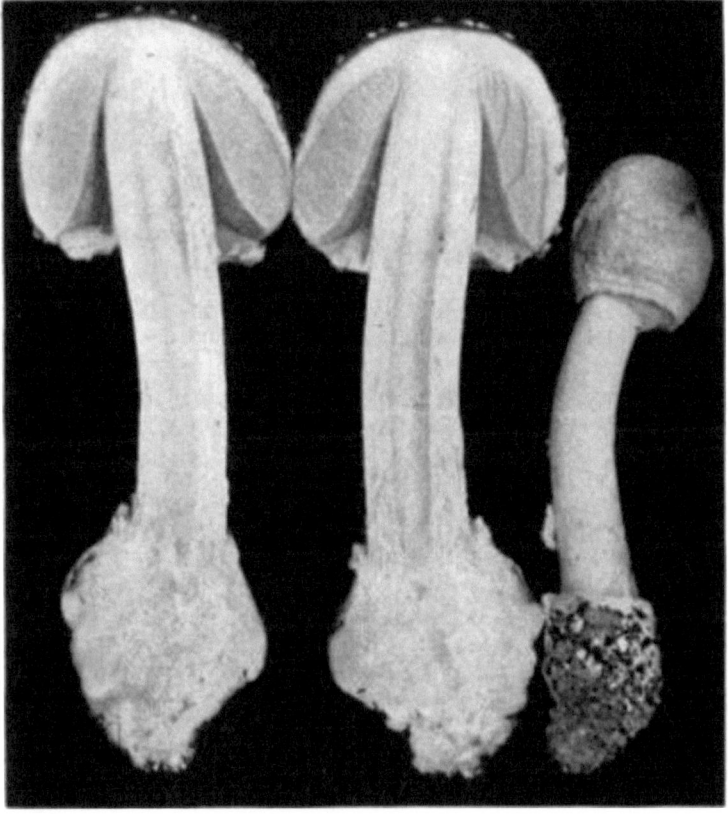

Figure 70.—Amanita cothurnata. Two plants in section showing clearly hollow stem, veil attachment, etc. (natural size). Copyright.

In *Amanita frostiana* the remains of the volva sometimes form a similar collar, but not so stout, on the base of the stem. The variations in *A. frostiana* where the stem, annulus and gills are white might suggest that there is a close relationship between *A. frostiana* and *A. cothurnata*, and that the latter is only a form of the former. From a careful study of the two plants growing side by side the [Pg 68] evidence is convincing that the two are distinct. *Amanita frostiana* occurs also at Blowing Rock, appearing earlier in the season than *A. cothurnata*, and also being contemporary with it. *A. frostiana* is more variable, not nearly so viscid, nor nearly so abundant, the stem is solid or stuffed, the annulus is more frail and evolved from the stem in a different manner. The volva does not leave such a constant and well defined roll where it separated on the stem transversely, and the pileus is yellow or orange. When *A. cothurnata* is yellowish at all it is a different tint of yellow and then only a tinge of yellow at the center. Albino or faded forms of *A. frostiana* might occur, but we would not expect them to appear at a definite season of the year in great abundance while the normal form, showing no intergrading specimens in the same locality, continued to appear in the same abundance and with the same characters as before. The dried plants of *A. cothurnata* are apt to become tinged with yellow on the gills, the upper part of the stem and upper part of the annulus during the processes of drying, but the pileus does not change in like manner, nor do these plants show traces of yellow on these parts when fresh. The spores are also decidedly different, though the shape and size do not differ to any great extent. In *A. frostiana* and the pale forms of the species the spores are nearly globose or oval, rarely with a tendency to become elliptical, but *the content is quite constantly finely granular*, while the spores of *A. cothurnata* are perhaps more constantly globose or nearly so, but the spore is *nearly filled with a highly refractive oil globule or "nucleus."* The pileus of [Pg 69] *A. frostiana* is also thinner than that of *A. cothurnata*. It is nearer, in some respects, to specimens of *Amanita pantherina* received from Bresadola, of Austria-Hungary.

Figure 71.—Amanita spreta. The two outside plants show the free limb of the volva lying close against the stem (natural size, often larger). Copyright.

Amanita spreta Pk. **Said to be Poisonous.**—According to Peck this species grows in open or bushy places. The specimens illustrated in Fig. 71 grew in sandy ground by the roadside near trees in the edge of an open field at Blowing Rock, N. C., and others were found in a grove. The plants are 10–15 cm. high, the caps 6–12 cm. broad, and the stems 8–12 mm. in thickness. The **pileus** is convex to expanded, gray or light drab, and darker on the center, or according to Dr. Peck it may be white. It is smooth, or with only a few remnants of the volva, striate on the margin, and 1—.5 cm. thick at the center. The **gills** are white, adnexed, that is they reach the stem by their upper angle. The **stem** is of the same color as the pileus, but somewhat lighter, white to light gray or light drab, cylindrical, not bulbous, hollow or stuffed. The **annulus** is thin and attached above the middle of the stem. The **volva** is sordid white, and sheathes the stem with a long free limb of 3–5 lobes. It splits at [Pg 70] the apex, but portions sometimes cling to the surface of the pileus.

Figure 71 is from plants (No. 3707, C. U.) collected at Blowing Rock, N. C., September, 1899.

Amanita cæsarea Scop. **Edible**, *but use great caution.*—This plant is known as the orange amanita, royal agaric, Cæsar's agaric, etc. It is one of the most beautiful of all the agarics, and is well distributed over the earth. With us it is more common in the Southern States. It occurs in the summer and early autumn in the woods. It is easily recognized by its usually large size, yellow or orange color of the cap, gills, stem and ring, and the prominent, white, sac-like volva at the base of the stem. It is usually 12–20 cm. high, the cap 5–10 cm. broad, and the stems 6–10 mm. in thickness, though it may exceed this size, and depauperate forms are met with which are much smaller.

The **pileus** is ovate to bell-shaped, convex, and finally more or less expanded, when the surface may be nearly flat or the center may be somewhat elevated or umbonate and the margin curved downward. The surface is smooth except at the margin, where it is prominently striate. The color varies from orange to reddish or yellow, usually the well developed and larger specimens have the deeper and richer colors, while the smaller specimens have the lighter colors, and the color is usually deeper on the center of the pileus. The **gills** are yellow, and free from the stem. The **stem** is hollow, even in young plants, when it may be stuffed with loose threads. It is often very floccose scaly below the annulus. It is cylindrical, only slightly enlarged below, where it is covered by the large, fleshy, sac-like white volva. The **annulus** is membranaceous, large, and hangs like a broad collar from the upper part of the stem. The stem and ring are orange or yellow, the depth of the color varying more with the size of the plant than is the case with the color of the cap. In small specimens the stem is often white, especially in depauperate specimens are the stem and annulus white, and even the gills are white when the volva may be so reduced as to make it difficult to distinguish the specimens from similar specimens of the poisonous fly agaric.

Plate 18, Figure 72. — Amanita cæsarea. Different stages of development (2/3 natural size). Cap, stem, gills, veil orange or yellow. Volva white. Copyright.

In the button stage the plant is ovate and the white color of the volva, which at this time entirely surrounds the plants, presents an appearance not unlike that of an egg. The volva splits open at the apex as the stem elongates. The veil is often connected by loose threads with the outer portion of the stem and as the pileus expands this is torn away, leaving coarse floccose scales on the stem. Some of the different stages in the opening of the plant are shown in Fig. 72. This illustration is taken from a photograph of plants (No. 3726, C. [Pg 71] U. herbarium) collected at Blowing Rock, N. C., September, 1899. The plant is said to be one of the best esculents, and has been prized as an article of food from ancient times. Great caution should be used in distinguishing it from the fly agaric and from other amanitas.

- PLATE 19.
- Fig. 1. — Amanita rubescens
- Fig. 2. — A. cæsarea.
- Copyright 1900.

Amanita rubescens Fr. **Edible**, *but use great caution.* — The reddish amanita, *Amanita rubescens*, is so called because of the sordid reddish color diffused over the entire plant, and especially because bruised portions quickly change to a reddish color. The plant is often quite large, from 12–20 cm. high, the cap 8–12 cm. broad and the stem 8–12 mm. in thickness, but it is sometimes much smaller. It occurs during the latter part of the summer and in early autumn, in woods and open places.

Figure 73.—Amanita rubescens. Plant partly expanded. Dull reddish brown, stains reddish when bruised; for other details see text (natural size). Copyright.

The **pileus** is oval to convex, and becoming expanded when old. It is smooth or faintly striate on the margin, and covered with numerous scattered, thin, floccose, grayish scales, forming remnants of the larger part of the volva or outer veil. The color of the cap varies correspondingly, but is always tinged more or less distinctly with pink, red, or brownish red hues. The **gills** are white or whitish and free from the stem. The **stem** is nearly cylindrical, tapering some [Pg 72] above, and with a prominent bulb which often tapers abruptly below. In addition to the suffused dull reddish color the stem is often stained with red, especially where handled or touched by some object. There are very few evidences of the volva on the stem since the volva is so floccose and torn into loose fragments, most of which remain on the surface of the cap. Sometimes a few of these loose fragments are seen on the upper portion of the bulb, but they are easily removed by handling or by rains. The **annulus** is membranous, broad, and fragile.

Since the plant has become well known it is regarded as excellent and wholesome for food and pleasant to the taste. In case of the larger specimens there should be no difficulty in distinguishing it from others by those who care to compare the descriptions closely with the fresh specimens. But as in all cases beginners should use extreme caution in eating plants they have not become thoroughly familiar with. Small specimens of this species sometimes show but little of the reddish color, and are therefore difficult to determine.

Figures 73 and 74 are from plants (No. 3727 C. U. herbarium) collected at Blowing Rock, N. C., during September, 1899.

Amanita solitaria Bull. **Edible**, *but use caution.*—The solitary amanita, like many other plants, is not always true to its name. While it often occurs solitary, it does occur sometimes in groups. It is one of the largest of the amanitas. Its large size, together with its chalky white or grayish white color, and ragged or shaggy appearance, makes it a striking object in the woods, or along roadsides in woods where it grows. Frequently parts of the cap, the entire stem and the gills are covered with a white, crumbly, floccose substance of a mealy consistency which often sticks to the hands or other objects. The plant ranges from 15–20 cm. or more high, the cap from 8–15 cm. broad, and the stems are 1–2 cm. or more in thickness.

In form the **pileus** ranges from nearly globose in the button stage, to hemispherical, convex and expanded, when quite old the margin becoming more or less elevated. It is covered either with flaky or floccose portions of the volva, or with more or less distinct conic white scales, especially toward the center. The conic scales are easily rubbed off in handling or are easily washed off by rains. Many of them are loosened and fall because of the tension produced by the expanding pileus on the surface of which they rest. These scales vary in size from quite small ones, appearing like granules, to those fewer in number and larger, 3 mm. high and nearly as broad at the base. In other cases the scales are harder and stouter and dark colored. These forms will be discussed after the description of the other parts of the plant.

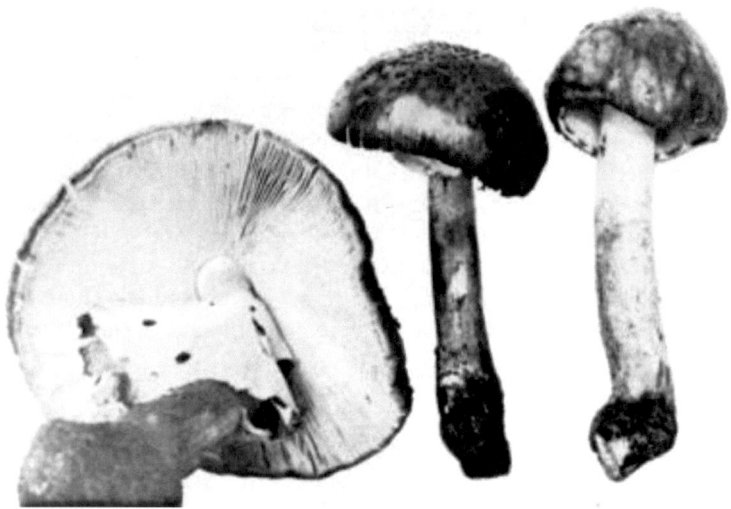

Plate 20, Figure 74.—Amanita rubescens. Under and side view. Dull reddish brown, stains reddish where bruised (3/4 natural size). Copyright.

[Pg 73]

Plate 21, Figure 75.—Amanita solitaria. Entirely white, or cap and scales sordid buff, dull brown, or grayish in some plants. For details see text (1/2 natural size). Copyright.

The **gills** are free, or are only attached by the upper inner angle; the edges are often floccose where they are torn from the slight union with the upper surface of the veil. The **stem** is cylindrical, solid or stuffed when old, enlarged usually below into a prominent bulb which then tapers into a more or less elongated root-like process, sometimes extending 5–10 cm. in the ground below the bulb. In rare cases the bulb is not present, but the cylindrical stem extends for a considerable distance into the ground. The **veil** is a very interesting part of the plant and the manner in which it forms and disappears as the cap expands is worth a careful study. This is well shown in Figs. 75, 76, from photographs of plants (No. 3731 C. U. herbarium) made at Blowing Rock, N. C., during September, 1899.

During the latter part of August and the first three weeks of September the plants were quite common in the mountain woods at Blowing Rock. In certain features there was close agreement in the case of all the specimens examined, especially in the long rooting character of the base of the stem. The veil and annulus were also quite constant in their characters, though sometimes a tendency was manifested to split up more irregularly than at other times. In the character of the warts of the pileus there was great variation, showing typical forms of *Amanita solitaria* and grading into forms which might be taken for typical *Amanita strobiliformis*. Especially is this so in the case of some of my specimens (No. 3733), where the scales are pyramidal, dark brown, surrounded by a sordid buff or grayish area, and these latter areas separated by narrow chinks whitish in color. The scales in this specimen are fixed quite firmly to the surface of the pileus. In other specimens (No. 3732) these hard scales remove quite easily, while in still another the pileus is almost smooth, even the floccose scales having been obliterated, while a very few of the hard angular warts are still present. In another half expanded plant (of No. 3732) the warts are pyramidal, 4–6 mm. long at the center of the pileus and rather closely imbricated, hard, and firmly joined to the surface of the cap. In Nos. 3733 and 3731 the spores measure 7–9 × 4–6 μ. In 3732 they are longer, varying from 7–11 μ.

The specimens with the long hard scales suggest *Amanita strobiliformis* Vittad., but the long rooting base of the stem does not agree with the description of that plant, but does clearly agree with *Amanita solitaria* Bull. A study of the variations in these plants suggests that *Amanita solitaria* and *strobiliformis* Vittad., represent only variations in a single species as Bulliard interpreted the species more than a century ago. Forms of the plant are also found which suggest that [Pg 74] *A. polypyramis* B. & C., collected in North Carolina, is but one of the variations of *A. solitaria*.

Figures 75, 76 show well certain stages in the development of this plant. The conical or pyramidal warts are formed in a very young stage of the plant by the primary separation of the outer part of the volva, and as the pileus expands more, and the cessation of growth of the outer veil proceeds inward, the scales become more widely separated at the apex and broader at the base. In some cases the volva is probably thinner than in others, and with the rapid expansion of the pileus in wet weather the scales would be smaller, or more floccose. But with different conditions, when it is not so wet, the plant expands less rapidly, the surface of the pileus becomes drier, the volva layer does not separate so readily and the fissures between the scales proceed deeper, and sometimes probably enter the surface of the pileus, so that the size of the warts is augmented. A similar state of things sometimes takes place on the base of the stem at the upper margin of the bulb, where the concentric fissures may extend to some distance in the stem, making the scales here more prominent in some specimens than in others. A similar variation in the character of the scales on the bulb of *Amanita muscaria* is sometimes presented.

The veil is often loosely attached to the edges of the gills, and so is stripped off from the stem quite early. Sometimes it is more strongly adherent to the stem, or portions of it may be, when it is very irregularly ruptured as it is peeled off from the stem, as shown in the plant near the left side in Fig. 75. The veil is very fragile and often tears a little distance from the margin of the cap, while the portion attached to the stem forms the annulus. This condition is shown in the case of three plants in Fig. 75. The plant is said to be edible.

AMANITOPSIS Roze.

This genus has white spores, and a volva, but the annulus and inner veil are wanting. In other respects it agrees with *Amanita*. It is considered as a sub-genus of *Amanita* by some.

Plate 22, Figure 76.—Amanita solitaria. Three plants, 3/4 natural size. Copyright.

Amanitopsis vaginata (Bull.) Roz. **Edible.**—The sheathed amanitopsis, *A. vaginata*, is a quite common and widely distributed plant in woods. It is well named since the prominent volva forms a large sheath to the cylindrical base of the stem. The plant occurs in several forms, a gray or mouse colored form, and a brownish or fulvous form, and sometimes nearly white. These forms are recognized [Pg 75] by some as varieties, and by others as species. The plants are 8–15 cm. high, the caps 3–7 cm. broad, and the stems 5–8 mm. in thickness.

Plate 23, Figure 77. — Amanitopsis vaginata. Tawny form (natural size). Copyright.

The **pileus** is from ovate to bell-shaped, then convex and expanded, smooth, rarely with fragments of the volva on the surface. The margin is thin and marked by deep furrows and ridges, so that it is deeply striate, or the terms sulcate or pectinate sulcate are used to express the character of the margin. The term pectinate sulcate is employed on account of a series of small elevations on the ridges, giving them a pectinate, or comb-like, appearance. The color varies from gray to mouse color, brown, or ochraceous brown. The flesh is white. The **gills** are white or nearly so, and free. The **spores** are globose, 7–10 μ in diameter. The **stem** is cylindrical, even, or slightly tapering upward, hollow or stuffed, not bulbous, smooth, or with mealy particles or prominent floccose scales. These scales are formed by the separation of the edges of the gills from the surface of the stem, to which they are closely applied before the pileus begins to expand. Threads of mycelium growing from the edge of the lamellæ and from the stem intermingle. When the pileus expands these are torn asunder, or by their pull tear up the outer surface of the stem. The **volva** forms a prominent sheath which is usually quite soft and easily collapses (Fig. 77).

The entire plant is very brittle and fragile. It is considered an excellent one for food. I often eat it raw when collecting.

Authors differ as to the number of species recognized in the plant as described above. Secretan recognized as many as ten species. The two prominent color forms are quite often recognized as two species, or by others as varieties; the gray or mouse colored form as *A. livida* Pers., and the tawny form as *A. spadicea* Pers. According to Fries and others the *livida* appears earlier in the season than *spadicea*, and this fact is recognized by some as entitling the two to specific rank. Plowright (Trans. Brit. Mycol. Soc., p. 40, 1897–98) points out that in European forms of *spadicea* there is a second volva inside the outer, and in *livida* there are "folds or wrinkles of considerable size on the inner surface of the volva." He thinks the two entitled to specific rank. At Ithaca and in the mountains of North Carolina I have found both forms appearing at the same season, and thus far have been unable to detect the differences noted by Plowright in the volva. But I have never found intergrading color forms, and have not yet satisfied myself as to whether or not the two should be entitled to specific rank.

Some of the other species of *Amanitopsis* found in this country [Pg 76] are **A. nivalis** Grev., an entirely white plant regarded by some as only a white form of **A. vaginata**. Another white plant is **A. volvata** Pk., which has elliptical spores, and is striate on the margin instead of sulcate.

Figure 78.—Amanitopsis farinosa. Cap grayish (natural size). Copyright.

Amanitopsis farinosa Schw.—The mealy agaric, or powdery amanita, is a pretty little species. It was first collected and described from North Carolina by de Schweinitz (Synop. fung. Car. No. 552, 1822), and the specimens illustrated in Fig. 78 were collected by me at Blowing Rock, N. C., during September, 1899. Peck has given in the 33rd Report N. Y. State Mus., p. 49, an excellent description of the plant, though it often exceeds somewhat the height given by him. It ranges from 5-8 or 10 cm. high, the cap from 2-3 cm. broad, and the stem 3-6 mm. in thickness.

The **pileus** is from subglobose to convex and expanded, becoming nearly plane or even depressed by the elevation of the margin in old specimens. The color is gray or grayish brown, or mouse colored. The pileus is thin, and deeply striate on the margin, covered with a grayish floccose, powdery or mealy substance, the remnant of the evanescent volva. This substance is denser at the center and is easily rubbed off. The **gills** are white and free from the stem. The [Pg 77] **spores** are subglobose and ovate to elliptical, 6-7 μ long. The **stem** is cylindrical, even, hollow or stuffed, whitish or gray and

131

very slightly enlarged at the base into a small rounded bulb which is quite constant and characteristic, and at first is covered on its upper margin by the floccose matter from the volva.

Plate 24, Figure 79.—Lepiota naucina. Entirely white (natural size).

At Blowing Rock the plants occurred in sandy soil by roadsides or in open woods. In habit it resembles strikingly forms of *Amanitopsis vaginata*, but the volva is entirely different (Fig. 78). Although *A. vaginata* was common in the same locality, I searched in vain for intermediate forms which I thought might be found. Sometimes the floccose matter would cling together more or less, and portions of it remained as patches on the lower part of the stem, while depauperate forms of *A. vaginata* would have a somewhat reduced volva, but in no case did I find intermediate stages between the two kinds of volva.

LEPIOTA Fr.

The genus *Lepiota* lacks a volva, but the veil is present forming a ring on the stem. The genus is closely related to *Amanita*, from which it differs in the absence of the volva, or perhaps more properly speaking in the fact that the universal veil is firmly connected (concrete with) with the pileus, and with the base of the stem, so that a volva is not formed. The gills are usually free from the stem, some being simply adnexed, but in some species connected with a collar near the stem. The stem is fleshy and is easily separable from the cap. A number of the species are edible. Peck, 35th Report N. Y. State Mus., p. 150–164, describes 18 species. Lloyd, Mycol. Notes, November, 1898, describes 9 species.

Lepiota naucina Fr. (*Lepiota naucinoides* Pk., *Annularia lævis* Krombh.) **Edible.**—The smooth lepiota, *L. naucina*, grows in lawns, in pastures and by roadsides, etc. It occurs during the latter part of summer and during autumn, being more abundant in September and early October. It is entirely white, or the cap is sometimes buff, and in age the gills become dirty pink in color. It is from 8–12 cm. high, the cap 5–10 cm. broad, and the stem 8–15 mm. in thickness.

The **pileus** is very fleshy, nearly globose, then convex to nearly expanded, smooth, or rarely the surface is broken into minute scales. The **gills** are first white, free from the stem, and in age assume a dull pink tinge. The **spores** are usually white in mass, but rarely when caught on white paper they show a faint pink tinge. The spores are elliptical to oval. The **stem** is nearly cylindrical, gradually enlarging below so that it is clavate, nearly hollow or stuffed with loose threads.

Figure 80.—Lepiota naucina.—Section of three plants, different ages.

[Pg 78] Since the plant occurs in the same situations as the *Agaricus campestris* it might be mistaken for it, especially for white forms. But of course no harm could come by eating it by mistake for the common mushroom, for it is valued just as highly for food by some who have eaten it. If one should look at the gills, however, they would not likely mistake it for the common mushroom because the gills become pink only when the plant is well expanded and quite old. There is much more danger in mistaking it for the white amanitas, *A. phalloides*, *A. verna*, or *A. virosa*, since the gills of these deadly plants are white, and they do sometimes grow in lawns and other grassy places where the smooth lepiota and the common mushroom grow. For this reason one should study the descriptions and illustrations of these amanitas given on preceding pages, and especially should the suggestions given there about care in collecting plants be followed, until one is so certainly familiar with the characters that the plants would be known "on sight."

Plate 25, Figure 81. — Lepiota procera. Grayish brown to reddish brown, gills and flesh white (3/4 natural size). Copyright.

The pink color of the gills of this lepiota has led certain students of the fungi into mistakes of another kind. This pink color of the gills has led some to place the plant among the rosy spored agarics in the genus *Annularia*, where it was named *Annularia lævis* by Krombholtz (vide Bresadola Funghi Mangerecci e velenosi, p. 29, [Pg 79] 1899). It fits the description of that plant exactly. The pink color of the gills, as well as the fact that the gills turn brownish when dry, has led to a confusion in some cases of the *Lepiota naucina* with the chalky agaric, *Agaricus cretaceus*. The external resemblance of the plants, as shown in various illustrations, is very striking, and in the chalky agaric the gills remain pink very late, only becoming brown when very old.

Lepiota procera Scop. **Edible.** — The parasol mushroom, *Lepiota procera*, grows in pastures, lawns, gardens, along roadsides, or in thin woods, or in gardens. It is a large and handsome plant and when expanded seems not inappropriately named. It is from 12–20 cm. or more high, the cap expands from 5–12 cm., while the stem is 4–7 mm. in thickness. It occurs during summer and in early autumn.

The **pileus** is oval, then bell-shaped, convex and nearly expanded, with usually a more or less prominent elevation (umbo) at the center. Sometimes it is depressed at the center. It is grayish brown or reddish brown in color on the surface and the flesh is whitish. As the cap expands the surface layer ceases to grow and is therefore cracked, first narrow chinks appearing, showing white or grayish threads underneath. As the cap becomes more expanded the brown surface is torn into scales, which give the cap a more or less shaggy appearance except on the umbo, where the color is more uniform. The torn surface of the pileus shows numerous radiating fibres, and it is soft and yielding to the touch. The **gills** are remote from the stem, broad and crowded. The **spores** are long, elliptical, 12–17 μ long. The **stem** is cylindrical, hollow, or stuffed, even, enlarged below into a prominent bulb, of the same color as the pileus, though paler, especially above the annulus. The surface is usually cracked into numerous small scales, the chinks between showing the white inner portion of the stem. The **ring** is stout, narrow, usually quite free from the stem, so that it can be moved up and down on the stem, and is called a movable ring.

Figure 81 is from plants (No. 3842, C. U. herbarium) collected in a garden at Blowing Rock, N. C., during September, 1899.

A closely related plant, *Lepiota rachodes* Vitt., has smaller spores, 9–12 × 7–9 μ. It is also edible, and by some considered only a variety of *L. procera*. It is rare in this country, but appears about Boston in considerable quantities "in or near greenhouses or in enriched soil out of doors," where it has the appearance of an introduced plant

(Webster, Rhodora, 1: 226, 1899). It is a much stouter plant than *L. procera*, the pileus usually depressed, much more coarsely scaly, and usually grows in dense clusters, while *L. procera* usually [Pg 80] occurs singly or scattered, is more slender, often umbonate. *L. rachodes* has a veil with a double edge, the edges more or less fringed. The veil is fixed to the stem until the plant is quite mature, when it becomes movable. The flesh of the plant on exposure to the air becomes a brownish orange tint.

Figure 82.—Lepiota americana. Scales and center of cap reddish or reddish brown. Entire plant turns reddish on drying (natural size). Copyright.

Lepiota morgani Pk.—This plant occurs from Ohio, southward and west. It grows in grassy places, especially in wet pastures. It is one of the largest of the lepiotas, ranging from 20–40 cm. high, the cap 20–30 cm. broad, and the stem about 2 cm. in thickness. The **pileus**, when fully expanded, is whitish, with large dark scales, especially toward the center. The **ring** is large, sometimes movable, and the **gills** and **spores** are greenish. Some report the plant as edible, while others say illness results from eating it.

Lepiota americana Pk. **Edible.**—This plant is widely distributed in the United States. The plants occur singly or are clustered, 6–12 [Pg 81] cm. high, the cap 4–10 cm. broad, and the stem 4–10 mm. in thickness. The cap is adorned with reddish or reddish brown scales except on the center, where the color is uniform because the surface is not broken up into scales. The flesh is white, but changes to reddish when cut or bruised, and the whole plant becomes reddish on drying.

Figure 82 is from plants (No. 2718, C. U. herbarium) collected at Ithaca.

The European plant, *L. badhami*, also reported in this country, changes to a brownish red. It is believed by some to be identical with *L. americana*.

Figure 83.—Lepiota cristata. Entirely white, but scales grayish or pinkish brown, stem often flesh color (natural size). Copyright.

Lepiota acutesquamosa Weinm.—This is a medium or small sized plant with a floccose pileus adorned with small, acute, erect scales, and has a loose, hairy or wooly veil which is often torn irregularly. The erect scales fall away from the pileus and leave little scars where they were attached.

Lepiota cristata A. & S. **Edible.** — The crested lepiota, *Lepiota cristata*, occurs in grassy places and borders of woods, in groves, etc., from May to September, and is widely distributed. The plant is small, 3–5 cm. high, the cap 1–4 cm. broad, and the stem 2–5 mm. in thickness. It grows in clusters or is scattered.

The **pileus** is ovate, bell-shaped, then convex and expanded, and thin. The surface is at first entirely dull reddish or reddish brown, but soon cracks into numerous scales of the same color arranged in a crested manner, more numerous between the margin and the center, and often arranged in a concentric manner. The center of the cap [Pg 82] often preserves the uniform reddish brown color because the pileus at this point does not expand so much and therefore the surface does not crack, while the margin often becomes white because of the disappearance of the brown covering here. The **gills** are free from the stem, narrow, crowded, and close to the stem. The **spores** are more or less angular, elongated, more narrowed at one end, and measure 5–8 × 3–4 μ. The **stem** is slender, cylindrical, hollow, whitish, smooth. The **ring** is small, white, and easily breaks up and disappears.

The characters of the plant are well shown in Fig. 83 from plants collected at Ithaca. *Lepiota angustana* Britz. is identical, and according to Morgan *L. miamensis* Morgan is a white form of *L. angustana*.

Lepiota asperula Atkinson. — This lepiota resembles *A. asper* in some respects, but it is smaller and the spores are much smaller, being very minute. The plant is 5–8 cm. high, the pileus 2–4 cm. broad, and the stem 4–6 mm. in thickness. It grows in leaf mould in the woods and has been found at Ithaca, N. Y., twice during July and September, 1897.

The **pileus** is convex and bell-shaped, becoming nearly or quite expanded. It is hair brown to olive brown in color. The surface is dry, made up of interwoven threads, and is adorned with numerous small, erect, pointed scales resembling in this respect *A. asper* Fr. The **gills** are white or yellowish, free, but rather close to the stem, narrow, often eroded on the edge, sometimes forked near the stem, and some of them arranged in pairs. The spores are oblong, smooth,

and very minute, measuring 5 × 2 μ. The **stem** is the same color as
the pileus, cylindrical, hollow, with loose threads in the cavity, enlarged into a rounded bulb below, minutely downy to pubescent.
The outer portion of the bulb is formed of intricately interwoven
threads, among which are entangled soil and humus particles. The
veil is white, silky, hairy, separating from the stem like a dense
cortina, the threads stretched both above and below as shown in
Fig. 84 from plants (No. 3157 C. U. herbarium), collected at Ithaca.

In some specimens, as the pileus expands, the spaces between the
pointed scales are torn, thus forming quite coarse scales which are
often arranged in more or less concentric rows, showing the yellow-
tinged flesh in the cracks, and the coarse scales bearing the fine
point at the center. A layer connecting the margin of the pileus with
the base of the stem and covered with fine brown points, sometimes
separates from the edge of the cap and the base of the stem, and
clings partly to the cortina and partly to the stem in much the same
way that portions of the volva cling to the stem of certain species of
Amanita, as seen in *A. velatipes* (Fig. 66). Sometimes [Pg 83] this is
left on the base of the stem and then resembles a short, free limb of a
volva, and suggests a species of *Amanita*. The scales, however, are
concrete with the pileus, and the species appears to show a closer
relationship with *Lepiota*.

Plate 26, Figure 84.—Lepiota asperula. Cap hair-brown to olive-brown, scales minute, pointed, gills and stem white (natural size). Copyright.

ARMILLARIA Fr.

In the genus *Armillaria* the inner veil which forms a ring on the stem is present. The stem is fibrous, or the outer portion cartilaginous in some species, and not easily separable from the substance of the pileus (continuous with the hymenophore), and the gills are attached to the stem, sinuate, or decurrent, spores white. Peck, 43rd Report N. Y. State Mus., p. 40–45, describes 6 species.

Some of the species resemble very closely certain species of *Amanita* or *Lepiota*, but can be distinguished by the firm continuity of the substance of the stem and cap.

Armillaria mellea Vahl. **Edible.**—This is one of the most common of the late summer and autumn fungi, and is widely distributed over the world. It grows about the bases of old stumps or dead trees, or from buried roots. Sometimes it is found attached to the living roots of trees. The plant occurs in tufts or clusters, several to many individuals growing together, the bases of their stems connected with a black rope-like strand from which they arise. The entire plant is often more or less honey colored, from which the plant gets its specific name. Its clustered habit, the usually prominent ring on the stems, and the sharp, blackish, erect scales which usually adorn the center of the cap, mark it as an easy plant to determine in most cases. The colors and markings, however, vary greatly, so that some of the forms are very puzzling. The plant varies in height from 10–15 cm., the cap from 5–10 cm. broad, and the stem 4–10 mm. in thickness.

The **pileus** is oval to convex and expanded, sometimes with a slight umbo or elevation at the center. The color varies from honey color to nearly white, or yellowish brown to dull reddish brown, usually darker on the center. In typical forms the pileus is adorned with pointed dark brown, or blackish, erect, scales especially abundant over the center, while the margin is often free from them, but may be marked with looser floccose, brownish, or yellowish scales. Sometimes there are no blackish pointed scales anywhere on the cap, only loose floccose colored scales, or in some forms the cap is entirely smooth. The margin in old specimens is often striate. The pileus is usually dry, but Webster cites an instance in which it was viscid in wet weather.

The **gills** are attached to the stem squarely (adnate) or they are [Pg 84] decurrent (extend downward on the stem), are white, or whitish, becoming in age more or less dingy or stained. The spores are rounded or elliptical, 6–9 μ. The **stem** is elastic, spongy within and sometimes hollow. It is smooth or often floccose scaly below the ring, sometimes with prominent transverse bands of a hairy substance. It is usually whitish near the upper end, but dull brown or reddish brown below the annulus, sometimes distinctly yellowish. The **veil** varies greatly also. It may be membranaceous and thin, or quite thick, or in other cases may be absent entirely. The **ring** of course varies in a corresponding manner. As shown in Fig. 85 it is quite thick, so that it appears double on the edge, where it broke away from the inner and outer surfaces of the margin of the cap. It is frequently fixed to the stem, that is, not movable, but when very thin and frail it often disappears.

The honey colored agaric is said by nearly all writers to be edible, though some condemn it. It is not one of the best since it is of rather tough consistency. It is a species of considerable economic importance and interest, since it is a parasite on certain coniferous trees, and perhaps also on certain of the broad-leaved trees. It attacks the roots of these trees, the mycelium making its way through the outer layer, and then it grows beneath the bark. Here it forms fan-like sheets of mycelium which advance along both away from the tree and towards the trunk. It disorganizes and breaks down the tissues of the root here, providing a space for a thicker growth of the mycelium as it becomes older. In places the mycelium forms rope-like strands, at first white in color, but later becoming dark brown and shining. These cords or strands, known as *rhizomorphs*, extend for long distances underneath the bark of the root. They are also found growing in the hollow trunks of trees sometimes. In time enough of the roots are injured to kill the tree, or the roots are so weakened that heavy winds will blow the trees over.

The fruiting plants always arise from these rhizomorphs, and by digging carefully around the bases of the stems one can find these cords with the stems attached, though the attachment is frail and the stems are easily separated from the cords. Often these cords grow for years without forming any fruit bodies. In this condition they are often found by stripping off the bark from dead and rotting logs in the woods. These cords were once supposed to be separate fungi, and they were known under the name *Rhizomorpha subcorticalis*.

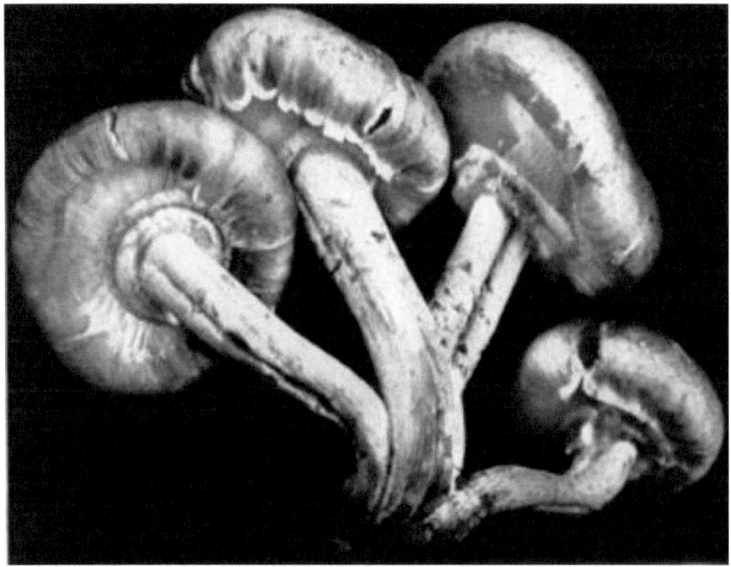

Plate 27, Figure 85. — Armillaria mellea. Showing double ring present in some large specimens; cap honey colored, scales minute, more numerous at center, blackish, often floccose, and sometimes wanting (3/4 natural size, often smaller). Copyright.

Armillaria aurantia Schaeff. (*Tricholoma peckii* Howe) **Suspected.** — This is a very pretty species and rare in the United States. The [Pg 85] plants are 6–8 cm. high, the cap 4–7 cm. broad, and the stem 6–8 mm. in thickness. It occurs in woods. It is known by its viscid pileus, the orange brown or ochraceous rufus color of the pileus and stem, and the color of the stem being confined to the superficial layer, which becomes torn into concentric floccose scales, forming numerous minute floccose irregular rings of color around the stem.

Figure 86.—Armillaria aurantia Schaeff. (=Tricholoma peckii Howe). Cap orange-brown or ochraceous rufus, viscid; floccose scales on stem same color (natural size). Copyright.

The **pileus** is convex to expanded, with an umbo, and the edge inrolled, fleshy, thin, viscid, ochraceous rufus (in specimens collected by myself), darker on the umbo, and minutely scaly from tufts of hairs, and the viscid cuticle easily peeling off. The **gills** are narrow, crowded, slightly adnexed, or many free, white, becoming brown discolored where bruised, and in drying brownish or rufus. The **spores** are minute, globose to ovoid, or rarely sub-elliptical when a little longer, with a prominent oil globule usually, 3–3.5×3–5 μ, sometimes a little longer when the elliptical forms are presented. The **stem** is straight or ascending, even, very floccose scaly as the pileus is unrolled from it, scales same color as the pileus, the scales running [Pg 86] transversely, being separated perhaps by the elongation of the stem so that numerous floccose rings are formed, showing the white flesh of the stem between. The upper part of the stem, that above the annulus, is white, but the upper part floccose.

Figure 87.—Tricholoma personatum. Entire plant grayish brown, tinged with lilac or purple, spores light ochraceous (natural size, often larger).

This plant has been long known in Europe. There is a rather poor figure of it in Schaeffer Table 37, and a better one in Gillet Champignons de France, Hymenomycetes, **1**, opposite page 76, but a very good one in Bresadola Funghi Mangerecci e Velenosi, Tavel 18, 1899. A good figure is also given by Barla, Les Champignons des Alpes—Maritimes, Pl. 19, Figs. 1–6. The plant was first reported from America in the 41st Report, State Museum, N. Y., p. 82, 1888, under the name *Tricholoma peckii* Howe, from the Catskill Mountains, N. Y. Figure 86 is from plants (No. 3991, C. U. herbarium) collected in the Blue Ridge mountains, at Blowing Rock, N. C., during September, 1899. The European and American description both ascribe a bitter taste to the flesh of the pileus, and it is regarded as suspicious.

There does not seem to be a well formed annulus, the veil only being present in a rather young stage, as the inrolled margin of the pileus is unrolling from the surface of the stem. It seems to be more in the form of a universal veil resembling the veil of some of the lepiotas. It shows a relationship with *Tricholoma* which possesses in typical forms a delicate veil present only in the young stage. Perhaps for this reason it was referred by Howe to *Tricholoma* as an undescribed species when it was named *T. peckii*. If its affinities should prove to be with *Tricholoma* rather than with *Armillaria*, it would then be known as *Tricholoma aurantium*.

[Pg 87] TRICHOLOMA Fr.

In the genus *Tricholoma* the volva and annulus are both wanting, the spores are white, and the gills are attached to the stem, but are more or less strongly notched or sinuate at the stem. Sometimes the notch is very slight. The stem is fleshy-fibrous, attached to the center of the pileus, and is usually short and stout. In some specimens when young there is a slight cobwebby veil which very soon disappears. The genus is a very large one. Some species are said to be poisonous and a few are known to be edible. Peck, 44th Report, N. Y. State Mus., pp. 38–64, describes 46 species.

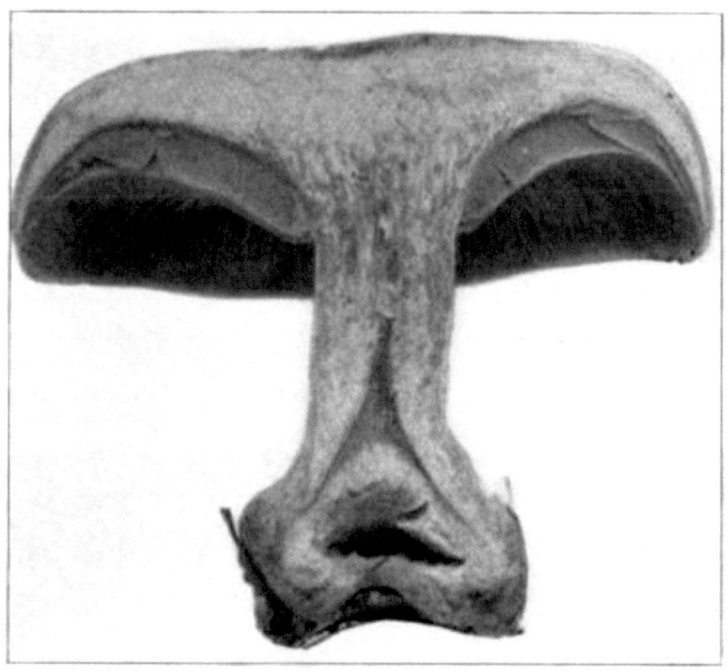

Figure 88.—Tricholoma personatum. Section (natural size).

Tricholoma personatum Fr. **Edible.**—This plant occurs during the autumn and persists up to the winter months. It grows on the ground in open places and in woods. The stem is short, usually 3–7 cm. long × 1–2 cm. in thickness, and the cap is from 5–10 cm. or more broad. The entire plant often has a lilac or purple tint.

The **pileus** is convex, expanded, moist, smooth, grayish to brownish tinged with lilac or purple, especially when young, fading out in age. When young the pileus is sometimes adorned with white mealy particles, and when old the margin may be more or less upturned and wavy. The **gills** are crowded, rounded next the stem, and nearly free but close to the stem, violet or lilac when young, changing to dull reddish brown when old. The **spores** when caught in mass are dull pink or salmon color. They measure 7–9 μ long. The **stem** is solid, fibrous, smooth, deep lilac when young and retaining the lilac color longer than the pileus. Sometimes the base is bulbous as in Fig. 87.

This plant is regarded by all writers as one of the best of the edible fungi. Sometimes the pileus is water soaked and then the [Pg 88] flavor is not so fine. The position of the plant is regarded as doubtful by some because of the more or less russety pink color of the spores when seen in mass, and the ease with which the gills separate from the pileus, characters which show its relationship to the genus *Paxillus*.

Tricholoma sejunctum Sowerb. **Edible.** — This plant occurs on the ground in rather open woods during late summer and in the autumn. It is 8–12 cm. high, the cap 5–8 cm. broad, and the stem 10–15 mm. in thickness.

Figure 89. — Tricholoma sejunctum. Cap light yellow, streaked with dark threads on the surface, viscid. Stem and gills white (natural size, often larger). Copyright.

The **pileus** is convex to expanded, umbonate, viscid when moist, light yellow in color and streaked with dark threads in the surface.

148

The flesh is white, and very fragile, differing in this respect from *T. equestre*, which it resembles in general form. The **gills** are broad, rather distant, broadly notched near the stem, and easily separating from the stem. The **stem** is solid, smooth and shining white. Figure 89 is from plants collected at Ithaca. It is said to be edible.

Plate 28, Figure 90.—Clitocybe candida. Entirely white (natural size). Copyright.

[Pg 89]

Plate 29, Figure 91. — Clitocybe candida. Under view of nearly lateral stemmed individual (natural size). Copyright.

CLITOCYBE Fr.

The volva and annulus are wanting in this genus, and the spores are white. The stem is elastic, spongy within, the outside being elastic or fibrous, so that the fibres hold together well when the stem is twisted or broken, as in *Tricholoma*. The stem does not separate readily from the pileus, but the rather strong fibres are continuous with the substance of the pileus. The gills are narrowed toward the stem, joined squarely or decurrent (running down on the stem), very rarely some of them notched at the stem while others of the same plant are decurrent. In one species at least (*C. laccata*, by some placed in the genus *Laccaria*) the gills are often strongly notched or sinuate. The cap is usually plane, depressed, or funnel-shaped, many of the species having the latter form. The plants grow chiefly on the ground, though a number of species occur on dead wood. The genus contains a very large number of species. Peck describes ten species in the 23rd Report, N. Y. State Mus., p. 76, et. seq., also 48th Report, p. 172, several species. Morgan, Jour. Cinn. Soc. Nat. Hist. **6**: 70–73, describes 12 species.

Clitocybe candida Bres. **Edible.**—This is one of the large species of the genus. It occurs in late autumn in Europe. It has been found on several occasions during late autumn at Ithaca, N. Y., on the ground in open woods, during wet weather. It occurs in clusters, though the specimens are usually not crowded. The stem is usually very short, 2–4 cm. long, and 2–3 cm. in thickness, while the cap is up to 10–18 cm. broad.

The **pileus** is sometimes regular, but often very irregular, and produced much more strongly on one side than on the other. It is convex, then expanded, the margin first incurved and finally wavy and often somewhat lobed. The color is white or light buff in age. The flesh is thick and white. The **gills** are white, stout, broad, somewhat decurrent, some adnate.

The taste is not unpleasant when raw, and when cooked it is agreeable. I have eaten it on several occasions. Figures 90, 91 are from plants (No. 4612 C. U. herbarium) collected at Ithaca.

Clitocybe laccata Scop. **Edible.**—This plant is a very common and widely distributed one, growing in woods, fields, roadsides and other waste places. It is usually quite easily recognized from the whitish scurfy cap, the pink or purplish gills, though the spores are white, from the gills being either decurrent, adnate, or more or less strongly notched, and the stem fibrous and whitish or of a pale pink color. When the plants are mature the pale red or pink gills appear [Pg 90] mealy from being covered with the numerous white spores.

The **pileus** is thin, convex or later expanded, of a watery appearance, nearly smooth or scurfy or slightly squamulose. The **spores** are rounded, and possess spine-like processes, or are prominently roughened. In the warty character of the spores this species differs from most of the species of the genus *Clitocybe*, and some writers place it in a different genus erected to accommodate the species of *Clitocybe* which have warty or spiny spores. The species with spiny spores are few. The genus in which this plant is placed by some is *Laccaria*, and then the plant is called *Laccaria laccata*. There are several other species of *Clitocybe* which are common and which one is apt to run across often, especially in the woods. These are of the funnel form type, the cap being more or less funnel-shaped. **Clitocybe infundibuliformis** Schaeffer is one of these. The cap, when mature, is pale red or tan color, fading out in age. It is 5–7 cm. high, and the cap 2–4 cm. broad. It is considered delicious. **Clitocybe cyathiformis**, as its name indicates, is similar in form, and occurs in woods. The pileus is of a darker color, dark brown or smoky in color.

Clitocybe illudens Schw. **Not Edible.** — This species is distributed through the Eastern United States and sometimes is very abundant. It occurs from July to October about the bases of old stumps, dead trees, or from underground roots. It is one of the large species, the cap being 15–20 cm. broad, the stem 12–20 cm. long, and 8–12 mm. in thickness. It occurs in large clusters, several or many joined at their bases. From the rich saffron yellow color of all parts of the plant, and especially by its strong phosphorescence, so evident in the dark, it is an easy plant to recognize. Because of its phosphorescence it is sometimes called "Jack-my-lantern."

The **pileus** is convex, then expanded, and depressed, sometimes with a small umbo, smooth, often irregular or eccentric from its crowded habit, and in age the margin of the pileus is wavy. The flesh is thick at the center and thin toward the margin. In old plants the color becomes sordid or brownish. The **gills** are broad, not crowded, decurrent, some extending for a considerable distance down on the stem while others for a less distance. The **stem** is solid, firm, smooth, and tapers toward the base.

While the plant is not a dangerously poisonous one, it has occasioned serious cases of illness, acting as a violent emetic, and of course should be avoided. Its phosphorescence has often been observed. Another and much smaller plant, widely distributed in this country as well as Europe, and belonging to another genus, is also phosphorescent. It is *Panus stipticus*, a small white plant with [Pg 91] a short lateral stem, growing on branches, stumps, trunks, etc. When freshly developed the phosphorescence is marked, but when the plants become old they often fail to show it.

Figure 92.—Clitocybe illudens. Entire plant rich saffron yellow, old plants become sordid brown sometimes; when fresh shows phosphorescence at night (2/3 natural size, often much larger). Copyright.

Clitocybe multiceps Peck. **Edible.**—This plant is not uncommon during late summer and autumn. It usually grows in large tufts of 10 to 30 or more individuals. The caps in such large clusters are often irregular from pressure. The plants are 6–12 cm. high, the caps 5–10 cm. broad, and the stems 8–15 mm. in thickness. The **pileus** is white or gray, brownish gray or buff, smooth, dry, the flesh [Pg 92] white. The **gills** are white, crowded, narrow at each end. The **spores** are smooth, globose, 5–7 μ in diameter. The stems are tough, fibrous, solid, tinged with the same color as cap. Fig. 93 is from plants (No. 5467, C. U. herbarium) collected at Ithaca, October 14, 1900.

COLLYBIA Fr.

In the genus *Collybia* the annulus and volva are both wanting, the spores are white, the gills are free or notched, or sinuate. The stem is either entirely cartilaginous or has a cartilaginous rind, while the central portion of the stem is fibrous, or fleshy, stuffed or fistulose. The pileus is fleshy and when the plants are young the margin of the pileus is incurved or inrolled, i. e., it does not lie straight against the stem as in *Mycena*.

Many of the species of *Collybia* are quite firm and will revive somewhat after drying when moistened, but they are not coriaceous as in *Marasmius*, nor do they revive so thoroughly. It is difficult, however, to draw the line between the two genera. Twenty-five of the New York species of Collybia are described by Peck in the 49th Report N. Y. State Mus., p. 32 et seq. Morgan describes twelve species in Jour. Cinn. Soc. Nat. Hist., 6: 70–73.

Collybia radicata Rehl. **Edible.**—This is one of the common and widely distributed species of the genus. It occurs on the ground in the woods or groves or borders of woods. It is quite easily recognized by the more or less flattened cap, the long striate stem somewhat enlarged below and then tapering off into a long, slender root-like process in the ground. It is from this "rooting" character that the plant gets its specific name. It is 10–20 cm. high, the cap 3–7 cm. broad, and the stem 4–8 mm. in thickness.

The **pileus** is fleshy, thin, convex to nearly plane, or even with the margin upturned in old plants, and the center sometimes umbonate. It is smooth, viscid when moist, and often with wrinkles on the surface which extend radially. The color varies from nearly white in some small specimens to grayish, grayish brown or umber. The flesh is white. The **gills** are white, broad, rather distant, adnexed, i. e., joined to the stem by the upper angle. The **spores** are elliptical and about 15 × 10 μ. The **stem** is the same color as the pileus though paler, and usually white above, tapers gradually above, is often striate or grooved, or sometimes only mealy. The long tapering "root" is often attached to some underground dead root. Fig. 94 is

154

from plants (No. 5641, C. U. herbarium) collected at Ithaca, August, 1900.

Plate 30, Figure 93.—Clitocybe multiceps. Plants white or gray to buff or grayish brown. (Three-fourths natural size.) Copyright.

Plate 31, Fig. 94.—Collybia radicata. Caps grayish-brown to grayish and white in some small forms. (Natural size.) Copyright.

Plate 32, Fig. 95.—Collybia velutipes. Cap yellowish or reddish yellow, viscid, gills white, stem dark brown, velvety hairy (natural size). Copyright.

Collybia velutipes Curt. **Edible.**—This is very common in woods or [Pg 93] groves during the autumn, on dead limbs or trunks, or from dead places in living ones. The plants are very viscid, and the stem, except in young plants, is velvety hairy with dark hairs. Figure 95 is from plants (No. 5430, C. U. herbarium) collected at Ithaca, October, 1900.

Collybia longipes Bull., is a closely related plant. It is much larger, has a velvety, to hairy, stem, and a much longer root-like process to the stem. It has been sometimes considered to be merely a variety of *C. radicata*, and may be only a large form of that species. I have found a few specimens in the Adirondack mountains, and one in the Blue Ridge mountains, which seem to belong to this species.

Collybia platyphylla Fr. **Edible.**—This is a much larger and stouter plant than *Collybia radicata*, though it is not so tall as the larger specimens of that species. It occurs on rotten logs or on the ground about rotten logs and stumps in the woods from June to September. It is 8–12 cm. high, the cap 10–15 cm. broad, and the stem about 2 cm. in thickness.

The **pileus** is convex becoming expanded, plane, and even the margin upturned in age. It is whitish, varying to grayish brown or dark brown, the center sometimes darker than the margin, as is usual in many plants. The surface of the pileus is often marked in radiating streaks by fine dark hairs. The **gills** are white, very broad, adnexed, and usually deeply and broadly notched next the stem. In age they are more or less broken and cracked. The **spores** are white, elliptical, 7–10 × 6–7 μ.

The plant resembles somewhat certain species of *Tricholoma* and care should be used in selecting it in order to avoid the suspected species of *Tricholoma*.

MYCENA Fr.

The genus *Mycena* is closely related to *Collybia*. The plants are usually smaller, many of them being of small size, the cap is usually bell-shaped, rarely umbilicate, but what is a more important character the margin of the cap in the young stage is straight as it is applied against the stem, and not at first incurved as it is in *Collybia*, when the gills and margin of the pileus lie against the stem. The stem is cartilaginous as in *Collybia*, and is usually hollow or fistulose. The gills are not decurrent, or only slightly so by a tooth-like process. Some of the species are apt to be confused with certain species of *Omphalia* in which the gills are but slightly decurrent, but in *Omphalia* the pileus is umbilicate in such species, while in *Mycena* it is blunt or umbonate. The spores are white. A large number of the plants grow on leaves and wood, few on the ground. Some of those which grow on leaves might be mistaken for species of *Marasmius*, but in [Pg 94] *Marasmius* the plants are of a tough consistency, and when dried will revive again if moistened with water.

Some of the plants have distinct odors, as alkaline, or the odor of radishes, and in collecting them notes should be made on all these characters which usually disappear in drying. A few of the plants exude a colored or watery juice when bruised, and should not be confounded with species of *Lactarius*.

Mycena galericulata Scop. **Edible.** — *Mycena galericulata* grows on dead logs, stumps, branches, etc., in woods. It is a very common and very widely distributed species. It occurs from late spring to autumn. The plants are clustered, many growing in a compact group, the hairy bases closely joined and the stems usually ascending. The plants are from 5–12 cm. high, the caps from 1–3 cm. broad, and the slender stems 2–3 mm. in thickness.

The **pileus** is conic to bell-shaped, sometimes umbonate, striate to near the center, and in color some shade of brown or gray, but variable. The **gills** are decurrent by a tooth, not crowded, connected by veins over the interspaces, white or flesh colored. The slender **stems** are firm, hollow, and hairy at the base.

Figure 96.—Mycena polygramma, long-stemmed form growing on ground (= M. prælonga Pk.). Cap dark brown with a leaden tint, striate on margin; stem finely and beautifully longitudinally striate (natural size). Copyright.

Mycena polygramma Bull.—This plant is very closely related to *M. galericulata*, and has the same habit. It might be easily mistaken

[Pg 95] for it. It is easily distinguished by its peculiar bright, shining, longitudinally striate to sulcate stem. It usually grows on wood, but does occur on the ground, when it often has a very long stem. In this condition it was described by Peck in the 23rd Report, N. Y. State Mus., p. 81, as *Mycena prælonga*, from plants collected in a sphagnum moor during the month of June. This form was also collected at Ithaca several times during late autumn in a woods near Ithaca, in 1898. The plants are from 12–20 cm. high, the cap 1–2 cm. broad, and the stem 2–3 mm. in thickness.

The **pileus** is first nearly cylindrical, then conic, becoming bell-shaped and finally nearly expanded, when it is umbonate. It is smooth, striate on the margin, of a dark brown color with a leaden tint. The **gills** are narrow, white, adnate and slightly decurrent on the stem by a tooth. The very long **stem** is smooth, but marked with parallel grooves too fine to show in the photograph, firm, hollow, somewhat paler than the pileus, usually tinged with red, and hairy at the base. Figure 96 is from plants (No. 3113 C. U. herbarium), collected in a woods near Ithaca in damp places among leaves. A number of the specimens collected were attacked by a parasitic mucor of the genus *Spinellus*. Two species, *S. fusiger* (Link.) van Tiegh., and *S. macrocarpus* (Corda) Karst., were found, sometimes both on the same plant. The long-stalked sporangia bristle in all directions from the cap.

Figure 97.—Mycena pura. Entire plant rose, rose purple, violet, or lilac. Striate on margin of pileus (natural size, often much larger).

Mycena pura Pers.—This plant is quite common and very widely distributed, and occurs in woods and grassy open places, during late summer and in the autumn. The entire plant is nearly of a uniform color, and the color varies from rose, to rose purple, violet, or lilac. Plants from the Blue Ridge mountains of North Carolina were chiefly rose purple, very young plants of a much deeper color (auricula purple of [Pg 96] Ridgeway), while those collected at Ithaca were violet. The plants vary from 5–8 cm. high, the cap 2–3 cm. broad, and the stem 2–4 mm. stout. The plants are scattered or somewhat clustered, sometimes occurring singly, and again many covering a small area of ground.

The **pileus** is thin, conic, bell-shaped to convex and nearly expanded, sometimes with a small umbo, smooth, and finely striate on the margin, in age the striæ sometimes rugulose from the upturning of the margin. Sometimes the pileus is rugose on the center.

The **gills** vary from white to violet, rose, etc., they are adnate to sinuate, and in age sometimes become free by breaking away from the stem. They are broad in the middle, connected by vein-like elevations over the surface, and sometimes wavy and crenate on the edge, the edge of the gills sometimes white. The **spores** are white, oblong, 2.5–3.5 × 6–7 μ, smooth. The **basidia** are cylindrical, 20–25 × 3–4 μ, four-spored. There are a few **cystidia** in the hymenium, colorless, thin walled, clavate, the portion above the hymenium cylindrical, and 30–40 × 10–12 μ.

The **stem** is sometimes white when young, but later becomes of the same color as the pileus, often a lighter shade above. It is straight, or ascending, cylindrical, even, smooth, hollow, with a few white threads at the base.

Sometimes on drying the pileus becomes deeper in color than when fresh. The gills also become deeper in color in drying, though the edge remains white if white when fresh. Figure 97 is from plants (No. 3946, C. U. herbarium) collected at Blowing Rock, N. C., in August, 1899. The plants are often considerably larger than shown in the figure.

Figure 98.—Mycena epipterygia. Cap viscid, grayish, often tinged with yellowish or reddish in age, gills white, sometimes tinged with blue or red, stem yellowish, or same color as cap (natural size). Copyright.

Mycena epipterygia Scop.—This pretty little species is quite readily distinguished by the gray, conic or bell-shaped cap, the long, hollow, slender stem, and the viscid pellicle or skin which is quite easily peeled off from the stem or cap when moist. It grows in woods or [Pg 97] grassy places, or among moss, etc., on the ground or on very rotten wood. The plants are from 5–10 cm. high, the cap 1–2 cm. broad, and the stem about 2 mm. in thickness. It is widely distributed in Europe, America, and other North temperate countries.

The **pileus** is viscid when moist, ovate to conic or campanulate, and later more or less expanded, obtuse, the margin striate, and sometimes minutely toothed. The usual color is grayish, but in age it often becomes reddish. The **gills** are decurrent by a small tooth, and quite variable in color, whitish, then gray, or tinged with blue or red.

The **stem** is very slender, flexuous, or straight, fistulose, tough, with soft hairs at the base, usually yellowish, sometimes the same color as the cap, and viscid like the cap when moist. Figure 98 is from plants (No. 4547, C. U. herbarium) collected at Ithaca in August, 1899.

Mycena vulgaris Pers. — This common and pretty species is easily recognized by its smoky or grayish color, the umbilicate pileus and very slimy stem. It grows on decaying leaves, sticks, etc., in woods. It occurs in clusters. The plants are small, 3–5 cm. high, the cap 4–7 mm. broad, and the stem about 1.5 mm. in thickness.

The **pileus** is thin, bell-shaped, then convex, and depressed at the center, with a papilla usually in the center, finely striate on the margin, and slightly viscid. The **gills** are white, thin, and finally decurrent, so that from the form of the cap and the decurrent gills the plant has much the appearance of an *Omphalia*. The **stem** is very viscid, grayish in color, often rooting at the base, and with white fibrils at the base, becoming hollow.

Figure 99 is from plants collected in woods near Ithaca, during August, 1899.

Figure 99.—Mycena vulgaris. Entirely white, center of cap gray-ish, entire plant very slimy when moist (natural size). Copyright.

Mycena acicula Schaeff.—This is one of the very small mycenas, and with the brilliant red pileus and [Pg 98] yellow gills and stem it makes a very pretty object growing on leaves, twigs, or rotten wood in the forest. It occurs during summer and autumn. It is 2–5 cm. high, the cap 2–4 mm. broad, and the stem is thread-like.

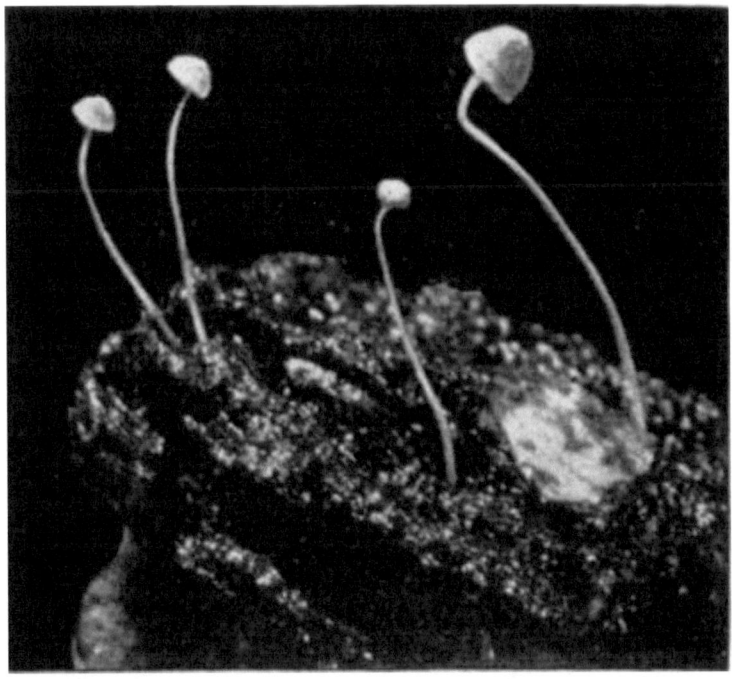

Figure 100.—Mycena acicula. Cap brilliant red, gills and stem yellowish (natural size). Copyright.

The **pileus** is very thin, membranaceous, bell-shaped, then convex, when the pointed apex appears as a small umbo. It is smooth, striate on the margin, and of a rich vermilion or orange color. The **gills** are rounded at the stem and adnexed, rather broad in the middle, distant, yellow, the edge white, or sometimes the gills are entirely white. The **stem** is very slender, with a root-like process entering the rotten wood, smooth except the hairs on the root-like process, yellow.

Figure 100 is from plants (No. 2780, C. U. herbarium) collected in a woods near Ithaca. It has been found here several times.

Mycena cyanothrix Atkinson.—This is a very pretty plant growing on rotting wood in clusters, often two or three joined at the base, the base of the stem inserted in the rotten wood for 1–2 cm., and the base is clothed with blue, hair-like threads. The plants are 6–9 cm.

high, the cap 1–2 cm. broad, and the stem not quite 2 mm. in diameter.

The **pileus** is ovate to convex, viscid when young. The color is bright blue when young, becoming pale and whitish in age, with a tendency to fuscous on the center. The cap is smooth and the margin finely striate. After the plants have dried the color is nearly uniform ochraceous or tawny. The **gills** are close, free, narrow, white, then grayish white, the edge finely toothed or fimbriate. The **spores** are globose, smooth, 6–9 μ. The **stem** is slender, hollow, faintly purple when young, becoming whitish or flesh color, flexuous, or nearly straight, even, often two united at the base into a root-like extension which enters the rotten wood. The base of the stem is covered with deep blue mycelium which retains its color in age, but disappears on drying after a time. Figure 101 is from plants (No. 2382, C. U. herbarium) collected at Ithaca, in woods, June 16, 1898.

Mycena hæmatopa Pers. — This is one of the species of *Mycena* with a red juice which exudes in drops where wounds occur on the plant. It is easily recognized by its dense cespitose habit, the deep blood [Pg 99] red juice, the hollow stem, and the crenate or denticulate sterile margin of the cap. Numbers of the plant occur usually in a single cluster, and their bases are closely joined and hairy. The stems are more or less ascending according to the position of the plant on the wood. The plants are 5–10 cm. high, the cap is 1–2.5 cm. broad, and the stem 2–3 mm. in thickness.

Figure 101.—Mycena cyanothrix. Cap viscid when young, blue, becoming pale and whitish in age, and fuscous in center; gills white; stem faintly purple when young, then flesh color or white, blue, clothed with blue hairs at base (natural size). Copyright.

The **pileus** is conic, then bell-shaped, and as the margin of the cap expands more appears umbonate, obtuse, smooth, even or somewhat striate on the margin. The color varies from whitish to flesh color, or dull red, and appears more or less saturated with a red juice. The thin margin extends a short distance beyond the ends of the gills, and the margin is then beautifully crenate. The **gills** are adnate, and often extend down on the stem a short distance by a little tooth. The **stem** is firm, sometimes smooth, sometimes with

169

[Pg 100] minute hairs, at the base with long hairs, hollow, in color the same as that of the pileus.

Figure 102.—Mycena hæmatopa. Dull red or flesh color, or whitish, a dull red juice exudes where broken or cut, margin of cap serrate with thin sterile flaps (natural size). Copyright.

The color varies somewhat, being darker in some plants than in others. In some plants the juice is more abundant and they bleed profusely when wounded, while in other cases there is but little of the juice, sometimes wounds only showing a change in color to a deep red without any free drops exuding. Figure 102 is from plants collected at Ithaca, in August, 1899. It is widely distributed in Europe and North America.

Mycena succosa Pk., another species of *Mycena* with a juice, occurs on very rotten wood in the woods. It is a small plant, dull white at first, but soon spotted with black, and turning black in handling or where bruised, and when dried. Wounds exude a "serum-like juice," and the wounds soon become black. It was described by Peck under *Collybia* in the 25th Report, p. 74.

OMPHALIA Fr.

The genus *Omphalia* is closely related to *Mycena* and *Collybia*. It differs from these mainly in the decurrent gills. In the small species of *Mycena* where the gills are slightly decurrent, the pileus is not umbilicate as it is in corresponding species of *Omphalia*. In some of the species of *Omphalia* the pileus is not umbilicate, but here the gills are plainly decurrent. The stem is cartilaginous.

[Pg 101]

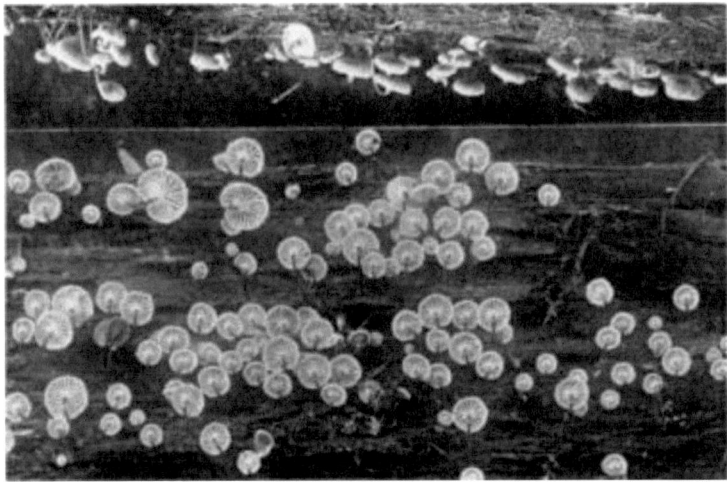

Plate 33, Figure 103.—Omphalia campanella. Watkin's Glen, N. Y., August, 1898. Caps dull reddish-yellow. Gills yellow. Stem brownish, hairy at base. (Natural size.) Copyright.

Omphalia campanella Batsch.—One of the most common and widely distributed species of the genus is the little bell-omphalia, *Omphalia campanella*. It occurs throughout the summer and autumn on dead or rotten logs, stumps, branches, etc., in woods. It is often clustered, large numbers covering a considerable surface of the decaying log. It is 1–3 cm. high, the cap 8–20 mm. broad, and the stem very slender.

The **pileus** is convex, umbilicate, faintly striate, dull reddish yellow, in damp weather with a watery appearance. The **gills** are narrow, yellow, connected by veins, strongly curved because of the form of the pileus, and then being decurrent on the stem. The **stem** is slender, often ascending, brownish hairy toward the base, and paler above.

Figure 104.—Omphalia epichysium. Entire plant smoky or dull gray in color (natural size). Copyright.

Omphalia epichysium Pers.—This plant occurs during the autumn in woods, growing usually on much decayed wood, or sometimes apparently on the ground. The smoky, or dull gray color of the entire plant, the depressed or funnel-shaped pileus, and short, slender stem serve to distinguish it. The cap is 2–4 cm. broad, the plant is 3–5 cm. high, and the stem 2–4 mm. in thickness.

The **pileus** is convex, becoming expanded, umbilicate or depressed at the center or nearly funnel-shaped, smooth, smoky or gray with a saturated watery appearance, light gray or nearly white when dry. The **gills** are narrow, crowded, or a little decurrent. The slender **stem** is smooth, hollow, equal. Figure 104 is from plants (No. 3373, [Pg 102] C. U. herbarium) collected in woods near Ithaca, N. Y., in the autumn of 1899.

PLEUROTUS Fr.

Figure 105.—Pleurotus ulmarius. Cap white, or with shades of yellow or brown near the center (natural size). Copyright.

The genus *Pleurotus* is usually recognized without difficulty among the fleshy, white-spored agarics, because of the eccentric (not quite in the center of the pileus) or lateral stem, or by the pileus being attached at one side in a more or less shelving position, or in some species where the upper side of the pileus lies directly against the wood on which the plant is growing, and is then said to be *resupinate*. The gills are either decurrent (extending downward) on the stem, or in some species they are rounded or notched at the junction with the stem. There is no annulus, though sometimes a veil, and the genus resembles both *Tricholoma* and *Clitocybe*, except for the position of the stem on the pileus. In *Tricholoma* and *Clitocybe* the stem is usually attached at the center, and the majority of the species grow on the ground, while the species of *Pleurotus* are especially characterized by growing on wood. Some species, at least, appear to grow from the ground, as in Pleurotus petaloides, which is sometimes found growing on buried roots or portions of decayed stumps which no longer show above ground. On the other hand species of *Clitocybe*, as in C. candida (Fig. 91), often have an eccentric stem. This presents to us one of the many difficulties which students, especially beginners, of this group of fungi meet, and also suggests how unsatisfactory any arrangement of genera as yet proposed is.

Pleurotus ulmarius Bull. **Edible.** — The elm pleurotus is so called [Pg 103] because it is often found growing on dead elm branches or trunks, or from wounds in living trees, but it is not confined to the elm. It is a large species, easily distinguished from the oyster agaric and the other related species by its long stem attached usually near the center of the cap, and by the gills being rounded or notched at their inner extremity. The cap is 5–12 cm. broad, the stem 5–10 cm. long, and 1–2 cm. in thickness.

Figure 106. — Pleurotus ulmarius. Under view and section (natural size). Copyright.

The **pileus** is convex, the margin incurved, then nearly expanded, smooth, firm, white or whitish, or with shades of yellow or brown on the center, and the flesh is white. The **gills** are broad, rather distant, [Pg 104] sinuate, white or nearly so. The **spores** are globose, 5–8 μ in diameter. The **stem** is firm, eccentric, usually curved because of its lateral attachment on the side of the tree, and the horizontal position of the pileus.

The elm pleurotus has been long known as an edible fungus, and is regarded as an excellent one for food on account of its flavor and because of its large size. It occurs abundantly during the late autumn, and at this season of the year is usually well protected from the attacks of insects. It occurs in the woods, or fields, more frequently on dead trees. On shade trees which have been severely pruned, and are nearly or quite dead, it sometimes appears at the wounds, where limbs have been removed, in great abundance. In the plants shown in Fig. 105 the stems are strongly curved because the weight of the cap bore the plant downward. Sometimes when the plant is growing directly on the upper side of a branch or log, the stem may be central.

Pleurotus ostreatus Jacq. **Edible.**—This plant is known as the oyster agaric, because the form of the plant sometimes suggests the outline of an oyster shell, as is seen in Fig. 107. It grows on dead trunks and branches, usually in crowded clusters, the caps often overlapping or imbricated. It is large, measuring 8–20 cm. or more broad.

The **pileus** is elongated and attached at one side by being sessile, or it is narrowed into a very short stem. It is broadest at the outer extremity, where it becomes quite thin toward the margin. It is more or less curved in outline as seen from the side, being depressed usually on the upper side near the point of attachment, and toward the margin convex and the margin incurved. The color is white, light gray, buff or dark gray, often becoming yellowish on drying. The **gills** are white, broad, not much crowded, and run down on the stem in long elevated lines resembling veins, which anastomose often in a reticulate fashion. The **spores** are white, oblong, 7–10 μ long. The **stem** when present is very short, and often hairy at the base.

The oyster agaric has long been known as an edible mushroom, but it is not ranked among the best, because, like most *Pleuroti*, it is rather tough, especially in age. It is well to select young plants. Figure 107 is from plants (No. 2097, C. U. herbarium) collected at Ithaca, N. Y.

Plate 34, Figure 107.—Pleurotus ostreatus. Under view showing decurrent and anastomosing gills on the stem. Cap white, light gray, buff, or dark gray in color. Spores white (natural size, often larger). Copyright.

Pleurotus sapidus Kalchb. **Edible.**—This plant usually grows in large clusters from dead trunks or branches or from dead portions of living trees. It grows on a number of different kinds of trees. The stems are often joined at the base, but sometimes the plants are scattered over a [Pg 105] portion of the branch or trunk. The cap is from 5–10 cm. broad. The plants occur from June to November.

Plate 35, Figure 108.—Pleurotus sapidus. Color of cap white, yellowish, gray, or brownish, with lilac tints sometimes. Spores lilac tinted in mass (1/2 natural size). Copyright.

The **pileus** is convex, the margin incurved when young, and more or less depressed in age, smooth, broadened toward the margin and tapering into the short stem, which is very short in some cases and elongated in others. Often the caps are quite irregular and the margin wavy, especially when old. It is quite firm, but the margin splits quite readily on being handled. The color varies greatly, white, yellowish, gray, or brownish and lilac tints. The flesh is white. The **stems** are usually attached to the pileus, at or near one edge. The **gills** are white, broad, not at all crowded, and extend down on the stem as in the oyster agaric. They are white or whitish, and as in the other related species are sometimes cracked, due probably to the tension brought to bear because of the expanding pileus. The **spores** are tinged with lilac when seen in mass, as when caught on paper. The color seems to be intensified after the spores have lain on the paper for a day or two.

It is very difficult to distinguish this species from the oyster agaric. The color of the spores seems to be the only distinguishing character, and this may not be constant. Peck suggests that it may only be a variety of the oyster agaric. I have found the plant growing from a dead spot on the base of a living oak tree. There was for several years a drive near this tree, and the wheels of vehicles cut into the roots of the tree on this side, and probably so injured it as to kill a portion and give this fungus and another one (*Polystictus pergamenus*) a start, and later they have slowly encroached on the side of the tree.

Figure 108 represents the plant (No. 3307, C. U. herbarium) from a dead maple trunk in a woods near Ithaca, collected during the autumn of 1899. This plant compares favorably with the oyster agaric as an edible one. Neither of these plants preserve as well as the elm pleurotus.

Pleurotus dryinus Pers. **Edible.**—*Pleurotus dryinus* represents a section of the genus in which the species are provided with a veil when young, but which disappears as the pileus expands. This species has been long known in Europe on trunks of oak, ash, willow, etc., and occurs there from September to October. It was collected near Ithaca, N. Y., in a beech woods along Six-mile creek, on October 24th, 1898, growing from a decayed knothole in the trunk of a living hickory tree, and again in a few days from a decayed stump. The pileus varies from 5–10 cm. broad, and the lateral or eccentric stem is 2–12 cm. long by 1–2 cm. in thickness, the length of the stem [Pg 106] depending on the depth of the insertion of the stem in a hollow portion of the trunk. The plant is white or whitish, and the substance is quite firm, drying quite hard.

The **pileus** is convex to expanded, more or less depressed in the center, the margin involute, and the surface at first floccose, becoming in age floccose scaly, since the surface breaks up into triangular scales more prominent in and near the center, smaller and inconspicuous toward the margin. The prevailing color is white, but in age the scales become cream color or buff (in European plants said to become fuscous). The pileus is either definitely lateral (Fig. 109) or eccentric when the stem is attached near the center as in Fig. 110. The **gills** are white, becoming tinged with yellow in age, decurrent (running down on the stem) in striæ for short distances, 4–5 mm. broad, not crowded. The **stem** is nearly central (Fig. 110), or definitely lateral (Fig. 109), the length varying according to conditions as stated above. It is firm, tough, fibrous. The **veil** is prominent in young and medium plants, floccose, tearing irregularly as the pileus expands.

Figure 110 is from plants (No. 2478a C. U. herbarium) growing from knothole in living hickory tree, and Fig. 109 from plants (No. 2478b) growing on a dead stump, near Ithaca.

According to the descriptions of *P. dryinus* as given by Persoon, and as followed by Fries and most later writers, the pileus is definitely lateral, and more or less dimidiate, while in *P. corticatus* Fr., the pileus is entire and the stem rather long and eccentric. Stevenson suggests (p. 166) that corticatus is perhaps too closely allied to dryinus. The plants in our Fig. 110 agree in all respects with *P. corticatus*, except that possibly the lamellæ do not anastomose on the stem as they are said to in *corticatus*. According to the usual descriptions *corticatus* is given as the larger species, while Fig. 109 of our plant, possessing the typical characters of *dryinus*, is the larger. The form of the pileus, the length and position of the stem, depends, as we know, to a large extent on the position of the plant on the tree. When growing from the upper side, so that there is room above for the expansion of the cap, the pileus is apt to be more regular, just as is the case in *Pleurotus ulmarius*, and the stem more nearly central. When the plant grows from a hollow place in the trunk as those shown in Fig. 110 did, then there is an opportunity for them to grow more or less erect, at least until they emerge from the hollow, and then the pileus is more nearly equal in its expansion and the stem is longer. Berkeley describes specimens of P. dryinus with long stems [Pg 107] growing from a hollow in an ash, and Stevenson (p. 167) reports the same condition.

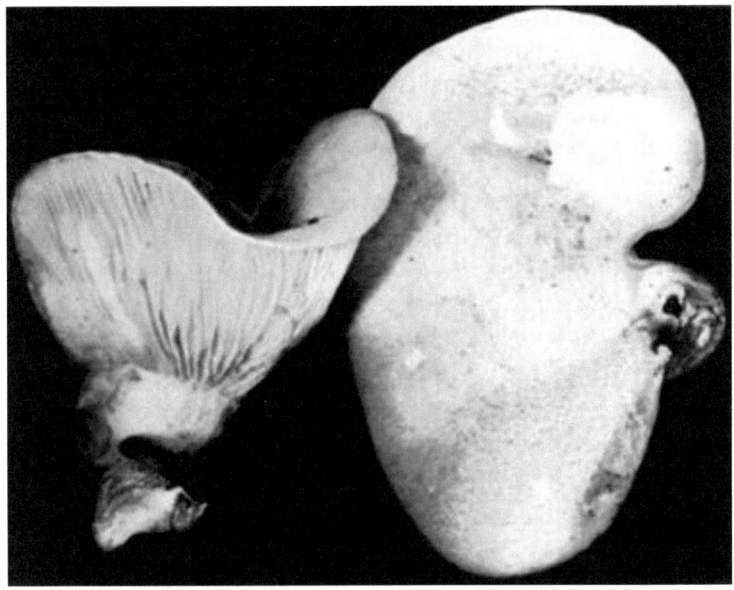

Plate 36, Figure 109.—Pleurotus dryinus. Side and upper view. Plant entirely white, scales sometimes buff or cream colored in age (natural size). Copyright.

Pleurotus sulfureoides Pk.—This rare species, first collected in the Catskill Mountains 1869, and described by Peck in the 23rd Report, N. Y. State Mus., p. 86, 1870, was found by me on two different occasions at Ithaca, N. Y., during the autumn of 1898, on rotting

180

logs, Ithaca Flats, and again in Enfield Gorge, six miles from Ithaca. The plants are from 5–8 cm. high, the cap 3–5 cm. broad, and the stem 5–7 mm. in thickness, and the entire plant is of a dull, or pale, yellow.

Plate 37, Figure 110.—Pleurotus dryinus, form corticatus. Entire plant white, scales cream or buff in age sometimes. The ruptured veil shows in the small plant below (natural size). Copyright.

The **pileus** is nearly regular, fleshy, thin toward the margin, convex, umbonate, smooth or with a few small scales. The **gills** are rather crowded, broad, rounded or notched at the stem, pale yellow. The **spores** are elliptical, 7–9 × 5–6 μ. The **stem** is ascending and curved, nearly or quite central in some specimens in its attachment to the pileus, whitish or yellowish, mealy or slightly tomentose at the apex.

Figure 111 is from plants (No. 2953, C. U. herbarium) on rotting log, Ithaca Flats, October, 1898.

Figure 111.—Pleurotus sulfureoides. Entire plant dull or pale yellow (natural size). Copyright.

Pleurotus petaloides Bull. **Edible.**—The petal-like agaric is so called from the fancied resemblance of the plant to the petal of a flower. The plant usually grows in a nearly upright or more or less ascending position, or when it grows from the side of a trunk it is somewhat shelving. It is somewhat spathulate in form, i. e., broad at the free [Pg 108] end and tapering downward into the short stem in a wedge-shaped manner, and varies from 2–10 cm. long and 1–5 cm. in breadth. It grows on fallen branches or trunks, on stumps, and often apparently from the ground, but in reality from underground roots or buried portions of decayed stumps, etc.

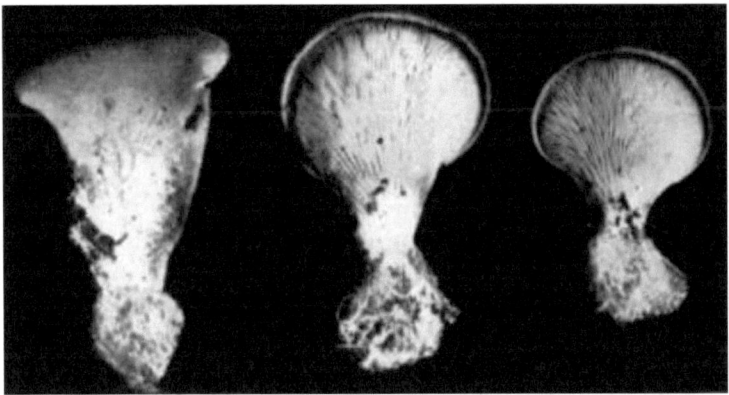

Figure 112.—Pleurotus petaloides. Color pale reddish brown or brown, sometimes entirely white; gills white (natural size). Copyright.

Figure 113.—Pleurotus petaloides. More irregular form than that shown in figure 112; color same as there described (natural size). Copyright.

The **pileus** varies from a regular wedge-shape to spathulate, or more or less irregularly petaloid, or conchoid forms, the extremes of size and form being shown in Figs. 112, 113. The margin is at first involute, finally fully expanded, and the upper surface is nearly plane or somewhat depressed. The color is often a pale reddish brown, or brown, and sometimes [Pg 109] pure white. The margin is sometimes marked with fine striations when moist. The upper portion near the union with the stem is sometimes tomentose, sometimes smooth. The **gills** are narrow, white, or yellowish, crowded and strongly decurrent. While the plant varies greatly in form and size, it is easily recognized by the presence of numerous short whitish **cystidia** in the hymenium, which bristle over the surface of the hymenium and under a pocket lens present a "fuzzy" appearance to the lamellæ. They are 70–80 × 10–12 μ. The spores are white.

Figures 112, 113 are from plants collected at Ithaca.

Pleurotus serotinus Schrad. This is an interesting plant and occurs during the autumn on dead trunks, branches, etc., in the woods. The stem is wanting, and the cap is shelving, dimidiate, reniform or suborbicular. The plants occur singly or are clustered and overlapping, about the same size and position as *Claudopus nidulans*, from which it is readily told by its white gills and spores. The color varies from dull yellow to brownish, often with shades of olive or green.

Pleurotus applicatus Batsch.—This is a pretty little species and usually occurs on much decayed wood, lying close to the ground so that it is usually directly on the under side of the log or branch. It does occur, however, on the side of the log when it is more or less shelving, because of the tendency of the pileus always to be more or less horizontal.

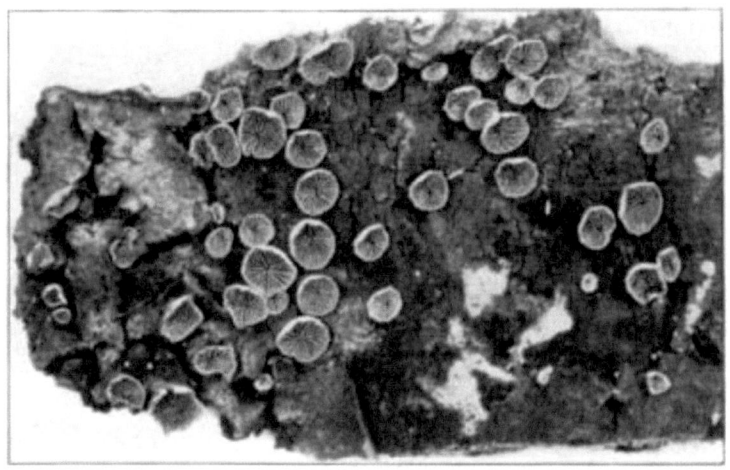

Figure 114.—Pleurotus applicatus. Color gray to dark bluish gray, or black with a bluish tinge (natural size). Copyright.

The **pileus** is 4–6 mm. broad, its upper surface closely applied to the wood or bark on which it is growing when it appears directly on the under side. The margin is sometimes free and involute. Sometimes it is attached only by the center of the pileus. There is then often a short process. When it grows on the side of the log it is attached laterally, or on the upper side of one margin, while the greater portion of the pileus is free and shelving. The surface is smooth or [Pg 110] somewhat hairy. The color varies from gray to dark bluish gray, or black with a bluish tinge. The **gills** are thick, broad in proportion to the size of the cap, distant, and are said by some to be paler than the pileus. In plants collected at Ithaca, the gills are often as dark as the pileus. The entire plant is rather tough, and revives after being dried if placed in water, resembling in this respect *Marasmius*, *Panus*, or *Trogia*, and it may be more nearly related to one of these. Figure 114 is from plants (No. 4599, C. U. herbarium) collected at Ithaca.

HYGROPHORUS Fries.

Figure 115.—Hygrophorus chrysodon. Entirely white with golden yellow granules on cap and stem (natural size). Copyright.

The genus *Hygrophorus* is one which presents some difficulties in the case of some of the species, especially to beginners, and plants need to be studied in the fresh condition to understand the most important character which separates it from certain of the other white-spored agarics. The substance of the pileus is continuous with that of the stem, that is, the stem is not easily separated from the cap at the point of junction, but is more or less tenacious. The gills may be adnexed, adnate, sinuate, or decurrent, but what is important they are usually rather distant, the edge is acute or sharp, and gradually thickened toward the junction with the cap, so that a section of the gill is more or less triangular. This is brought about by the fact that the substance of the cap extends downward into the gill between the laminæ or surfaces of the gill. But the most important character for determining the genus is the fact that the surfaces of the gills become rather of a waxy consistency at maturity, so that they appear to be full of a watery substance though they do not bleed, [Pg 111] and the surface of the gill can be rather easily removed, leaving the projecting line of the *trama*. This is more marked in some species than in others. The waxy consistency of the gills then, with the gills acute at the edge, broad at the point of attachment to the pileus, and the gills being rather widely separated are the important characters in determining the species which belong to this genus. The nearest related genus is Cantharellus, which, how-

186

ever, has blunt and forked gills. A number of the plants are brilliantly colored.

Plate 38, Figure 116. — Hygrophorus eburneus. Entirely white, slimy (natural size). Copyright.

Hygrophorus chrysodon (Batsch.) Fries. **Edible.** — This plant has about the same range as *Hygrophorus eburneus*, though it is said to be rare. It is a very pretty plant and one quite easily recognised because of the uniform white ground color of the entire plant when fresh, and the numerous golden floccules or squamules scattered over the cap and the stem. The name *chrysodon* means golden tooth, and refers to these numerous golden flecks on the plant. A form of the plant, variety *leucodon*, is said to occur in which these granules are white. The plant is 4–7 cm. high, the cap 4–7 cm. broad, and the stem 6–10 mm. in thickness. The plants grow on the ground in the woods, or rather open places, during late summer and autumn.

The **pileus** is convex, then expanded, the margin strongly involute when young, and unrolling as the cap expands, very viscid, so that particles of dirt and portions of leaves, etc., cling to it in drying. The golden or light yellow granules on the surface are rather numerous near the margin of the pileus, but are scattered over the entire surface. On the margin they sometimes stand in concentric rows close together. The **gills** are white, distant, decurrent, 3–6 mm. broad, white, somewhat yellowish in age and in drying, and connected by veins. The **spores** white, oval to ovate, the longer ones approaching elliptical, 6–10 × 5–6 µ.

The **stem** is soft, spongy within, nearly equal, white, the yellowish granules scattered over the surface, but more numerous toward the apex, where they are often arranged in the form of a ring. When the plant is young these yellow granules or squamules on the stem and the upper surface of the inrolled margin of the pileus meet, forming a continuous layer in the form of a veil, which becomes spread out in the form of separated granules as the pileus expands, and no free collar is left on the stem.

Figure 115 is from plants (No. 3108, C. U. herbarium) collected in October, 1898, in woods, and by roadsides, Ithaca, N. Y.

Hygrophorus eburneus (Bulliard) Fries. **Edible.** — This plant is widely distributed in Europe and America. It is entirely white, of medium size, very viscid or glutinous, being entirely covered with a coating [Pg 112] of gluten, which makes it very slippery in handling. The odor is mild and not unpleasant like that of a closely related species, *H. cossus*. The plants are 6–15 cm. high, the cap is from 3–8 cm. broad, and the stem 3–8 mm. in thickness. It grows on the ground in woods, or in open grassy places.

The **pileus** is fleshy, moderately thick, sometimes thin, convex to expanded, the margin uneven or sometimes wavy, smooth, and shining. When young the margin of the cap is incurved. The **gills** are strongly decurrent, distant, with vein-like elevations near the stem. **Spores** rather long, oval, 6–10 × 5–6 µ, granular. The **stem** varies in length, it is spongy to stuffed within, sometimes hollow and tapers below. The slime which envelops the plant is sometimes so abundant as to form a veil covering the entire plant and extending across from the margin of the cap to the stem, covering the gills. As the plant dries this disappears, and does not leave an annulus on the stem.

Figure 117.—Hygrophorus fuligineus. Cap and stem dull reddish brown or smoky brown, very viscid when moist; gills white (natural size). Copyright.

Figure 116 is from a photograph of plants (No. 2534, C. U. herbarium) collected in Enfield Gorge near Ithaca, N. Y., Nov. 5th, 1898. [Pg 113]

Hygrophorus fuligineus Frost. **Edible.**—The smoky hygrophorus was described in the 35th Report of the N. Y. State Museum, p. 134. It is an American plant, and was first collected at West Albany, during the month of November. It is one of the largest species of the genus, and grows on the ground in woods, in late autumn. The plants are 5–10 cm. high, the cap from 3–10 cm. broad, and the stem 1–2 cm. in thickness. The large size of the plant together with the smoky, brown, viscid cap aid in the recognition of the plant.

The **pileus** is convex, becoming expanded, smooth, very viscid, dull reddish brown or smoky brown, darker on the center; the margin of the pileus is even in young specimens, becoming irregular in others; and in age often elevated more or less. The **gills** are broad, distant, usually decurrent, often connected by veins, white, with yellowish tinge in drying. The **spores** oval to elliptical, 8–12 × 5–7 μ. The **stem** is stout, sometimes ascending, equal, or enlarged in the middle, or tapering toward the base, solid, viscid like the pileus, usually white, sometimes tinged with the same color as pileus, somewhat yellowish tinged in drying.

Figure 117 is from plants (No. 2546, C. U. herbarium) collected in Enfield Gorge near Ithaca, Nov. 5, 1898.

Hygrophorus pratensis (Pers.) Fr. **Edible.** — This hygrophorus grows on the ground in pastures, old fields, or in waste places, or in thin and open woods, from mid-summer to late autumn. The plants are 3–5 cm. high, the cap 2–5 cm. or more broad, and the stem 6–12 mm. in thickness. The cap being thick at the center, and the stem being usually stouter at the apex, often gives to the plant a shape like that of a top.

The **pileus** is hemispherical, then convex, then nearly or quite expanded, white, or with various shades of yellow or tawny, or buff, not viscid, often cracking in dry weather. Flesh very thick at the center, thinner at the margin. The flesh is firm and white. The **gills** are stout, distant, long decurrent, white or yellowish, and arcuate when the margin of the pileus is incurved in the young state, then ascending as the pileus takes the shape of an inverted cone. The **gills** are connected across the interspaces by vein-like folds, or elevations. The **spores** are nearly globose to ovate or nearly elliptical, white, 6–8 × 5–6 μ. The **stem** is smooth, firm outside and spongy within, tapering downward.

Hygrophorus miniatus Fr. The vermilion hygrophorus is a very common plant in the woods during the summer. The cap and stem are bright red, sometimes vermilion. The gills are yellow and often tinged with red. The gills are adnate or sinuate. The plant is a [Pg 114] small one but often abundant, and measures from 3–5 cm. high, and the cap 2–4 cm. broad. **Hygrophorus coccineus** (Schaeff.) Fr., is a somewhat larger plant and with a scarlet cap, which becomes yellowish in age, and the gills are adnate. **Hygrophorus conicus** (Scop.) Fr., is another bright red plant with a remarkable conical pileus, and the gills are annexed to free.

Hygrophorus psittacinus Fr., is a remarkably pretty plant, the cap being from bell-shaped to expanded, umbilicate, striate, and covered with a greenish slime. It occurs in woods and open places. The prevailing color is yellow, tinged with green, but it varies greatly, sometimes yellow, red, white, etc., but nearly always is marked by the presence of the greenish slime, the color of this disappearing as the plant dries. It occurs in pastures, open woods, etc., from mid-summer to autumn.

Hygrophorus hypothejus Fr., is another very variable plant in color as well as in size, varying from yellow, orange, reddish, sometimes paler, usually first grayish when covered with the olive colored slime. The gills are decurrent, white, then yellow. It occurs in autumn.

LACTARIUS Fr.

The genus *Lactarius* is easily distinguished from nearly all the other agarics by the presence of a milky or colored juice which exudes from wounded, cut, or broken places on the fresh plant. There are a few of the species of the genus *Mycena* which exude a watery or colored juice where wounded, but these are easily told from *Lactarius* because of their small size, more slender habit, and bell-shaped cap. By careful observation of these characters it is quite an easy matter to tell whether or not the plant at hand is a *Lactarius*. In addition to the presence of this juice or milk as it is commonly termed, the entire plant while firm is quite brittle, especially the gills. There are groups of rounded or vesiculose cells intermingled with thread-like cells in the substance of the cap. This latter character can only be seen on examination with the microscope. The brittleness of the plant as well as the presence of these groups of vesiculose cells is shared by the genus *Russula*, which is at once separated from *Lactarius* by the absence of a juice which exudes in drops.

In determining the species it is a very important thing to know the taste of the juice or of the fresh plant, whether it is peppery, or bitter, or mild, that is, tasteless. If one is careful not to swallow any of the juice or flesh of the plant no harm results from tasting any of the plants, provided they are not tasted too often during a short time, beyond the unpleasant sensation resulting from tasting some [Pg 115] of the very "hot" kinds. It is important also to know the color of the milk when it first exudes from wounds and if it changes color on exposure to the air. These tests of the plant should be made of course while it is fresh. The spores are white, globose or nearly so in all species, and usually covered with minute spiny processes.

There are a large number of species. Peck, 38th Report, N. Y. State Mus., pp. 111–133, describes 40 American species.

Figure 118.—Lactarius corrugis. Showing corrugated cap, and white milk exuding. Dark tawny brown, gills orange brown (natural size, often larger). Copyright.

Lactarius volemus Fr. **Edible.**—This species is by some termed the orange brown lactarius because of its usual color. It was probably termed *Lactarius volemus* because of the voluminous quantity of milk which exudes where the plant is broken or bruised, though it is not the only species having this character. In fresh, young plants, a mere crack or bruise will set loose quantities of the milky juice which drops rapidly from the plant. The plant is about the size of *Lactarius deliciosus* and occurs in damp woods, where it grows in considerable abundance from July to September, several usually [Pg 116] growing near each other. The **pileus** is convex, then expanded, often with a small elevation (umbo) at the center, or sometimes plane, and when old a little depressed in the center, smooth or somewhat wrinkled. The cap is dull orange or tawny, the shade of color being lighter in some plants and darker in others. The flesh is white and quite firm. The **gills** are white, often tinged with the same color as the pileus, but much lighter; they are adnate or slightly decurrent. The **stem** is usually short, but varies from 3–10 × 1–2 cm. It is colored like the pileus, but a lighter shade.

The milk is white, abundant, mild, not unpleasant to the taste, but sticky as it dries. This plant has also long been known as one of the excellent mushrooms for food both in Europe and America. Peck states that there are several plants which resemble *Lactarius volemus* in color and in the milk, but that no harm could come from eating them. There is one with a more reddish brown pileus, *Lactarius rufus*, found sparingly in the woods, but which has a very peppery taste. It is said by some to be poisonous.

Lactarius corrugis Pk. **Edible.** — This species occurs with *Lactarius volemus* and very closely resembles it, but it is of a darker color, and the pileus is more often marked by prominent wrinkles, from which character the plant has derived its specific name. It is perhaps a little stouter plant than *L. volemus*, and with a thicker cap. The surface of the **pileus** seems to be covered with a very fine velvety tomentum which glistens as the cap is turned in the light. The **gills** are much darker than in *L. volemus*. The plants are usually clearly separated on account of these characters, yet there are occasionally light colored forms of *L. corrugis* which are difficult to distinguish from dark forms of *L. volemus*, and this fact has aroused the suspicion that *corrugis* is only a form of *volemus*.

The milk is very abundant and in every respect agrees with that of *L. volemus*. I do not know that any one has tested *L. corrugis* for food. But since it is so closely related to *L. volemus* I tested it during the summer of 1899 in the North Carolina mountains. I consider it excellent. The methods of cooking there were rather primitive. It was sliced and fried with butter and salt. It should be well cooked, for when not well done the partially raw taste is not pleasant. The plant was very abundant in the woods, and for three weeks an abundance was served twice a day for a table of twelve persons. The only disagreeable feature about it is the sticky character of the milk, which adheres in quantity to the hands and becomes black. This makes the preparation of the plant for the broiler a rather unpleasant task. [Pg 117]

Figure 118 is from plants (No. 3910, C. U. herbarium) collected in the woods at Blowing Rock, during September, 1899. Just before the exposure was made to get the photograph several of the plants were wounded with a pin to cause the drops of milk to exude, as is well shown in the illustration.

The dark color of the lamellæ in *L. corrugis* is due to the number of brown cystidia or setæ, in the hymenium, which project above the surface of the gills, and they are especially abundant on the edge of the gills. These setæ are long fusoid, 80–120 × 10–12 μ. The variations in the color of the gills, in some plants the gills being much darker than in others, is due to the variations either in the number of these setæ or to the variation in their color. Where the cystidia are fewer in number or are lighter in color the lamellæ are lighter colored. Typical forms of *Lactarius volemus* have similar setæ, but they are very pale in color and not so abundant over the surface of the gills. In the darker forms of *L. volemus* the setæ are more abundant and darker in color, approaching those found in *L. corrugis*. These facts, supported by the variation in the color of the pileus in the two species and the variations in the rugosities of the pileus, seem to indicate that the two species are very closely related.

Figure 119.—Lactarius lignyotus. Cap and stem sooty, cap wrinkled, gills white, then tinged with ochre (natural size, sometimes larger). Copyright.

Lactarius lignyotus Fr. — This is known as the sooty lactarius and occurs in woods along with the smoky lactarius. It is distinguished from the latter by the dark brown color of the pileus and by the presence usually of rugose wrinkles over the center of the cap. In size it agrees with the smoky lactarius. [Pg 118]

The **pileus** is convex, then plane, or somewhat depressed in the center, dry, sometimes with a small umbo, dark brown or sooty (chocolate to seal brown as given in Ridgeway's nomenclature of colors), covered with a very fine tomentum which has the appearance of a bloom. The margin of the cap, especially in old plants, is somewhat wavy or plicate as in *Lactarius fuliginosus*. The **gills** are moderately crowded when young, becoming distant in older plants, white, then cream color or yellow, changing to reddish or salmon color where bruised. The **spores** are yellowish in mass, faintly so under the microscope, globose, strongly echinulate, 6–10 μ. The taste is mild, or sometimes slowly and slightly acrid. The plants from North Carolina showed distinctly the change to reddish or salmon color when the gills were bruised, and the taste was noted as mild.

Figure 119 is from plants (No. 3864, C. U. herbarium) collected in the Blue Ridge Mountains, at Blowing Rock, N. C., September, 1899.

Lactarius fuliginosus Fr. — The smoky or dingy lactarius occurs in woods and open grassy places. It is widely distributed. The plants are 4–7 cm. high, the cap 3–5 cm. broad, and the stem 6–10 mm. in thickness. The light smoky color of the cap and stem, the dull yellowish white color of the gills, and in old plants the wavy margin of the cap make it comparatively easy to recognize the species.

Figure 120.—Lactarius fuliginosus. Cap and stem smoky, cap usually not wrinkled; gills white, then light ochre, distant (natural size). Copyright.

The **pileus** is thin, at first firm, becoming soft, convex, then plane and often somewhat depressed in the center, usually even, dry, the margin in old plants crenately wavy, dull gray or smoky gray in color, with a fine down or tomentum. The **gills** are adnate, distant, more so in old plants, white, then yellowish, sometimes changing to salmon color or reddish where bruised. The **spores** are yellowish in mass, faintly yellow [Pg 119] under the microscope, strongly echinulate or tuberculate, globose, 6–10 μ. The **stem** is usually paler than the pileus, firm, stuffed. The milk is white, slowly acrid to the taste.

Figure 120 is from plants (No. 3867, C. U. herbarium) collected at Blowing Rock, N. C., during September, 1899.

Lactarius gerardii Pk.—This plant was described by Dr. Peck in the 26th Report, N. Y. State Mus., p. 65, and in the 28th Rept. p. 129. According to the descriptions it differs from *Lactarius fuliginosus* only in the spores being white, the gills more distant, and the taste being constantly mild. Since the taste in *L. fuliginosus* is sometimes mild, or slowly acrid, and the lamellæ in the older plants are more distant, the spores sometimes only tinged with yellow, there does not seem to be a very marked difference between the two species. In fact all three of these species, *fuliginosus*, *lignyotus* and *gerardii*, seem to be very closely related. Forms of *fuliginosus* approach *lignyotus* in color, and the **pileus** sometimes is rugose wrinkled, while in *lignyotus* pale forms occur, and the pileus is not always rugose wrinkled. The color of the bruised lamellæ is the same in the two last species and sometimes the change in color is not marked.

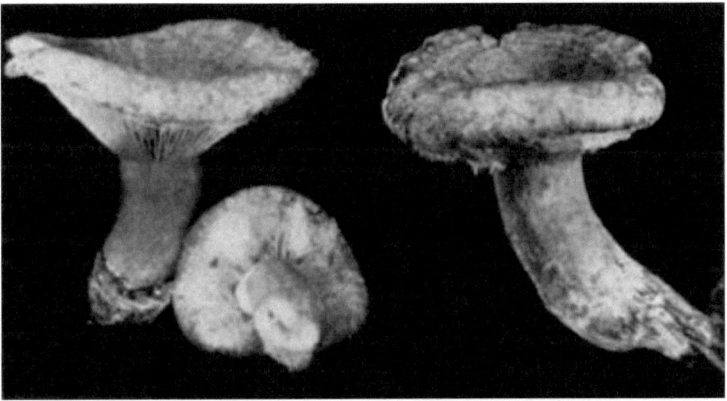

Figure 121.—Lactarius torminosus. Cap ochraceous and pink hues, with zones of darker color, margin of cap wooly (natural size, often much larger). Copyright.

Lactarius torminosus (Schaeff.) Fr.—This plant is widely distributed in Europe, Asia, as well as in America. It is easily recognised by the uneven mixture of pink and ochraceous colors, and the very hairy or tomentose margin of the cap. The plants are 5–10 cm. high, the cap about the same breadth, and the stem 1–2 cm. in thickness.

It occurs in woods on the ground during late summer and autumn. [Pg 120]

The **pileus** is convex, depressed in the center, and the margin strongly incurved when young, the abundant hairs on the margin forming an apparent veil at this time which covers up the gills. The upper surface of the pileus is smooth, or sometimes more or less covered with a tomentum similar to that on the margin. The color is an admixture of ochraceous and pink hues, sometimes with concentric zones of darker shades. The **gills** are crowded, narrow, whitish, with a tinge of yellowish flesh color. The **stem** is cylindrical, even, hollow, whitish.

The milk is white, unchangeable, acrid to the taste. Figure 121, left hand plants, is from plants (No. 3911, C. U. herbarium) collected in the Blue Ridge Mountains, N. C., in September, 1899, and the right hand plant (No. 2960, C. U. herbarium) collected at Ithaca, N. Y.

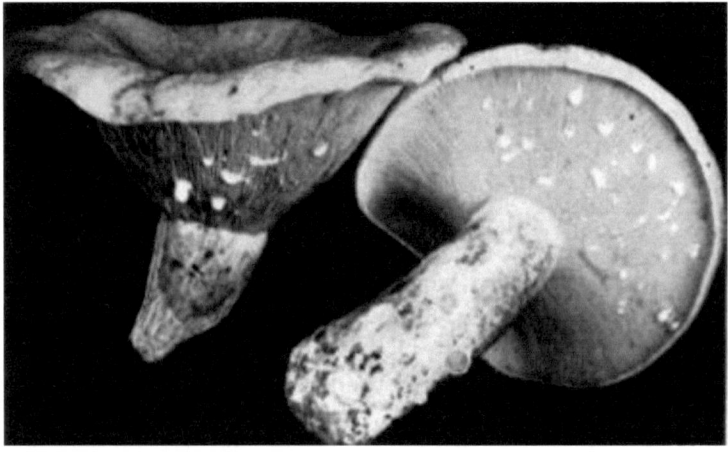

Figure 122.—Lactarius piperatus. Entirely white, milk very peppery (natural size, often larger). Copyright.

Lactarius piperatus (Scop.) Fr.—This species is very hot and peppery to the taste, is of medium size, entirely white, depressed at the center, or funnel-shaped, with a short stem, and very narrow and crowded gills, and abundant white milk. The plants are 3–7 cm. high, the cap 8–12 cm. broad, and the stem 1–2 cm. in thickness. It grows in woods on the ground and is quite common, sometimes very common in late summer and autumn.

The **pileus** is fleshy, thick, firm, convex, umbilicate, and then depressed in the center, becoming finally more or less funnel-shaped by the elevation of the margin. It is white, smooth when young, in age sometimes becoming sordid and somewhat roughened. The **gills** [Pg 121] are white, very narrow, very much crowded, and some of them forked, arcuate and then ascending because of the funnel-shaped pileus. The **spores** are *smooth*, oval, with a small point, 5–7 × 4–5 μ. The **stem** is equal or tapering below, short, solid.

The milk is white, unchangeable, very acrid to the taste and abundant. The plant is reported as edible. A closely related species is *L. pergamenus* (Swartz) Fr., which resembles it very closely, but has a longer, stuffed stem, and thinner, more pliant pileus, which is more frequently irregular and eccentric, and not at first umbilicate. Figure 122 is from plants (No. 3887, C. U. herbarium) collected at Blowing Rock, N. C., during September, 1899.

Figure 123.—Lactarius resimus. Entire plant white, in age scales on cap dull ochraceous (natural size). Copyright.

Lactarius resimus Fr.?—This plant is very common in the woods bordering a sphagnum moor at Malloryville, N. Y., ten miles from Ithaca, during July to September. I have found it at this place every summer for the past three years. It occurs also in the woods of the damp ravines in the vicinity of Ithaca. It was also abundant [Pg 122] in the Blue Ridge Mountains of North Carolina, during September, 1899. The plants are large, the caps 10–15 cm. broad, the stem 5–8 cm. long, and 2–3 cm. in thickness.

The **pileus** is convex, umbilicate, then depressed and more or less funnel-shaped in age, white, in the center roughened with fibrous scales as the plant ages, the scales becoming quite stout in old plants. The scales are tinged with dull ochraceous or are light brownish in the older plants. The ochre colored scales are sometimes evident over the entire cap, even in young plants. In young plants the margin is strongly involute or inrolled, and a loose but thick veil of interwoven threads extends from the surface of the roll to the stem. This disappears as the margin of the cap unrolls with the expanding pileus. The margin of the pileus is often sterile, that is, it extends beyond the ends of the gills. The **gills** are white, stout, and broad, decurrent, some of them forked near the stem. When bruised, the gills after several hours become ochraceous brown. The spores are subglobose, minutely spiny, 8–12 μ. The **stem** is solid, cylindrical, minutely tomentose, spongy within when old.

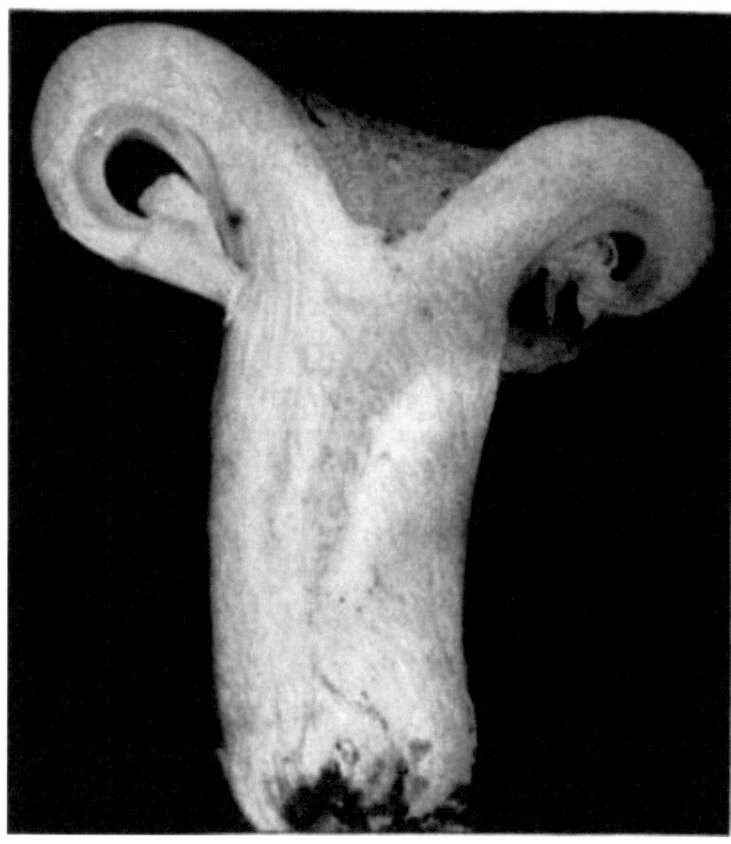

Figure 124.—Lactarius resimus. Section of young plant showing inrolled margin of cap, and the veil (natural size). Copyright.

The taste is very acrid, and the white milk not changing to yellow. While the milk does not change to yellow, broken portions of the plant slowly change to flesh color, then ochraceous brown. Figures 123, 124 are from plants collected in one of the damp gorges near Ithaca, during September, 1896. The forked gills, the strongly inrolled margin of the cap and veil of the young plants are well shown in the illustration.

Lactarius chrysorrheus Fr.—This is a common and widely distributed species, from small to medium size. The plants are 5–8 cm.

high, the cap 5–10 cm. broad, and the stem 1–1.5 cm. in thickness. It [Pg 123] grows in woods and groves during late summer and autumn.

The **pileus** is fleshy, of medium thickness, convex and depressed in the center from the young condition, and as the pileus expands the margin becomes more and more upturned and the depression deeper, so that eventually it is more or less broadly funnel-form. The color varies from white to flesh color, tinged with yellow sometimes in spots, and marked usually with faint zones of brighter yellow. The zones are sometimes very indistinct or entirely wanting. The **gills** are crowded, white then yellow, where bruised becoming yellowish, then dull reddish. The **stem** is equal or tapering below, hollow or stuffed, paler than the pileus, smooth (sometimes pitted as shown in the Fig. 125).

Figure 125.—Lactarius chrysorrheus. Cap white or flesh color, often tinged with yellowish, and with darker zones (natural size). Copyright.

The plant is acrid to the taste, the milk white changing to citron yellow on exposure. Figure 125 is from plants (No. 3875, C. U. herbarium) collected in the Blue Ridge Mountains at Blowing Rock, N. C., September, 1899. The species was quite abundant in this locality during August and September, in chestnut groves, mixed woods, and borders of woods.

Lactarius deliciosus (L.) Fr. **Edible.** — *Lactarius deliciosus* grows in damp woods, is widely distributed and sometimes is quite common. It occurs from July to October. It is one of the medium or large sized species, being 3–10 cm. high, the cap 5–12 cm. broad, and the [Pg 124] stem 1–2 cm. in thickness. It is easily recognized by its orange color and the concentric zones of light and dark orange around on the pileus, and by the orange milk which is exuded where wounded.

The **pileus** is first convex, then slightly depressed in the center, becoming more expanded, and finally more or less funnel-shaped by the elevation of the margin. It is usually more or less orange in color or mottled with varying shades, and with concentric bands of a deeper color. The **gills** are yellowish orange often with darker spots. The **stem** is of the same color as the pileus but paler, sometimes with darker spots. The flesh of the plant is white, shaded with orange. In old plants the color fades out somewhat and becomes unevenly tinged with green, and bruised places become green. Peck states that when fresh the plant often has a slight acrid taste.

Being a widely distributed and not uncommon plant, and one so readily recognized, it has long been known in the old world as well as here. All writers on these subjects concur in recommending it for food, some pronouncing it excellent, some the most delicious known. Its name suggests the estimation in which it was held when christened.

Lactarius chelidonium Pk. **Edible.** — This pretty little *Lactarius* was described by Peck in the 24th Report, N. Y. State Mus., p. 74. It is closely allied to *Lactarius deliciosus*, from which it is said to differ in its "more narrow lamellæ, differently colored milk, smaller spores." The plant is about 5 cm. high, the cap about 5 cm. broad, and the stem 1–1.5 cm. in thickness.

The **pileus** is fleshy, firm, convex and depressed in the center, smooth, slightly viscid when moist, "of a grayish green color with blue and yellow tints, and a few narrow zones on the margin." The **gills** are crowded, narrow, some of them forked at the base, and sometimes joining to form reticulations. The **spores** are yellowish. The short **stem** is nearly equal, smooth, hollow, and the same color as the pileus.

The taste is mild, the milk not abundant, and of a yellowish color, "resembling the juice of Celandine or the liquid secreted from the mouth of grasshoppers." Wounds on the plant are first of the color of the milk, changing on exposure to blue, and finally to green. The plant occurs during late summer and in the autumn in woods. Peck reported it first from Saratoga, N. Y. It has been found elsewhere in the State, and it has probably quite a wide distribution. I found it during September, 1899, in the Blue Ridge Mountains of N. C. Figure 1, plate 39, is from some of the water color drawings made by Mr. Franklin R. Rathbun.

- PLATE 39.
- Fig. 1.—Lactarius deliciosus.
- Fig. 2.—L. chelidonium.
- Fig. 3.—L. indigo.
- Copyright 1900.

[Pg 125] **Lactarius indigo** (Schw.) Fr. — The indigo blue lactarius is a very striking and easily recognized plant because of the rich indigo blue color so predominant in the entire plant. It is not very abundant, but is widely distributed in North America. The plant is 5–7 cm. high, the cap 5–12 cm. broad, and the stem is 1–2 cm. in thickness. The plants occur during late summer and in the autumn.

The **pileus** when young is umbilicate, the margin involute, and in age the margin becomes elevated and then the pileus is more or less funnel-shaped. The indigo blue color is deeply seated, and the surface of the pileus has a silvery gray appearance through which the indigo blue color is seen. The surface is marked by concentric zones of a darker shade. In age the color is apt to be less uniformly distributed, it is paler, and the zones are fainter. The *gills* are crowded, and when bruised, or in age, the indigo blue color changes somewhat to greenish. The milk is dark blue.

RUSSULA Pers.

The species of *Russula* are very characteristic, and the genus is easily recognized in most cases after a little experience. In the very brittle texture of the plants the genus resembles *Lactarius*, and many of them are more brittle than the species of this genus. A section of the pileus shows under the microscope a similar vesicular condition, that is the grouping of large rounded cells together, with threads between. But the species of *Russula* are at once separated from those of *Lactarius* by the absence of a juice which exudes in drops from bruised parts of *Lactarius*. While some of the species are white and others have dull or sombre colors, many of the species of *Russula* have bright, or even brilliant colors, as red, purple, violet, pink, blue, yellow, green. In determining many of the species, however, it is necessary to know the taste, whether mild, bitter, acrid, etc., and in this respect the genus again resembles *Lactarius*. The color of the gills as well as the color of the spores in mass should also be determined. The genus is quite a large one, and the American species are not well known, the genus being a difficult one. In Jour. Mycolog., **5**: 58–64, 1889, the characters of the tribes of Russula with descriptions of 25 species are quoted from Stevenson, with notes on their distribution in N. A. by MacAdam.

Russula alutacea Fr. **Edible.**—This handsome *Russula* differs from the others described here in the color of the gills and spores. The plant is common and occurs in mixed woods during the summer and early autumn. It is 5–10 cm. high, the cap 5–12 cm. broad, and the stem 1.5–2.5 cm. in thickness. [Pg 126]

The **pileus** is fleshy, oval to bell-shaped, becoming plane, and sometimes umbilicate. It is red or blood red in color, sometimes purple, and becoming pale in age, especially at the center. It is viscid when moist, the margin thin and striate-tuberculate. The **gills** are free from the stem, stout, broad, first white, becoming yellow, and in age ochraceous. The gills are all of the same length, not crowded, and they are connected by vein-like elevations over the surface. The **stem** is stout, solid, even, white, portions of the stem are red, sometimes purple.

The taste is mild, and the plant is regarded as one of the very good ones for food.

Russula lepida Fr. **Edible.**—This elegant *Russula* occurs in birch woods or in mixed woods during late summer and autumn. It is 5–8 cm. high, the cap 6–8 cm. broad, and the stem 1–2 cm. in thickness.

The **pileus** is fleshy, convex, then expanded, obtuse, not shining, deep red, becoming pale in age, often whitish at the center, silky, in age the surface cracking, the margin blunt and not striate. The **gills** are rounded next the stem, thick, rather crowded, and sometimes forked, white, sometimes red on the edge near the margin of the pileus. The gills are often connected by vein-like elevations over the surface. The **stem** is equal, white or rose color. The taste is mild.

Russula virescens (Schaeff.) Fr. **Edible.**—This plant grows on the ground in woods or in grassy places in groves from July to September. The stem is short, 2–7 cm. long × 1–2 cm. thick, and the cap is 5–10 cm. broad. The plant is well known by the green color of the pileus and by the surface of the pileus being separated into numerous, quite regular, somewhat angular areas or patches, where the green color is more pronounced.

The **pileus** is first rounded, then convex and expanded, and when old somewhat depressed in the center. It is quite firm, dry, greenish, and the surface with numerous angular floccose areas or patches of usually a deeper green. Sometimes the pileus is said to be tinged with yellow. The **gills** are adnate, nearly free from the stem, and crowded. The **stem** is white and firm.

The greenish Russula, *Russula virescens*, like a number of other plants, has long been recommended for food, both in Europe and in this country. There are several species of *Russula* in which the pileus is green, but this species is readily distinguished from them by the greenish floccose patches on the surface of the pileus. **Russula furcata** is a common species in similar situations, with forked gills, and the cap very variable in color, sometimes reddish, purple, purple [Pg 127] brown, or in one form green. I know of the *Russula furcata* having been eaten in rather small quantities, and while in this case no harm resulted the taste was not agreeable.

- PLATE 40.
- Fig. 1.—Russula virescens.
- Fig. 2.—R. alutacea.
- Fig. 3.—R. lepida.
- Fig. 4.—R. emetica.

- Fig. 5.—Yellow Russula.
  - Fig. 6.—R. adusta.
  - Copyright 1900.

Russula fragilis (Pers.) Fr.—This plant is very common in damp woods, or during wet weather from July to September. It is a small plant and very fragile, as its name suggests, much more so than most other species. It is 2–4 cm. high, the cap 2–5 cm. broad, and the stem about 1 cm. in thickness.

The **pileus** is convex, sometimes slightly umbonate, then plane, and in age somewhat depressed. The cuticle peels off very easily. The color is often a bright red, or pink, sometimes purple or violet, and becomes paler in age. It is somewhat viscid when moist, and the margin is very thin and strongly striate and tuberculate, i. e., the ridges between the marginal furrows are tuberculate. The **gills** are lightly adnexed, thin, crowded, broad, all of the same length, white. The **stem** is usually white, sometimes more or less pink colored, spongy within, becoming hollow. The taste is very acrid.

Russula emetica Fr. **Poisonous.**—This *Russula* has a very wide distribution and occurs on the ground in woods or open places during summer and autumn. It is a beautiful species and very fragile. The plants are 5–10 cm. high, the cap 5–10 cm. broad, and the stem 1–2 cm. in thickness. The **pileus** is oval to bell-shaped when young, becoming plane, and in age depressed. It is smooth, shining, the margin furrowed and tuberculate. The color is from pink or rosy when young to dark red when older, and fading to tawny or sometimes yellowish in age. The cuticle is easily separable as in *R. fragilis*, the flesh white, but reddish just beneath the cuticle. The **gills** are nearly free, broad, not crowded, white. The stem is stout, spongy within, white or reddish, fragile when old.

The plant is very acrid to the taste and is said to be poisonous, and to act as an emetic.

Russula adusta (Pers.) Fr.—This plant occurs on the ground in woods during late summer and in autumn. It is 3–6 cm. high, the cap 5–15 cm. broad, and the stem is 1–1.5 cm. in thickness.

211

The **pileus** is fleshy, firm, convex, depressed at the center, and when old more or less funnel-shaped from the upturning of the margin, which is at first incurved and smooth. It varies from white to gray and smoky color. The **gills** are adnate, or decurrent, thin, crowded, of unequal lengths, white, then becoming dark. The **stem** is colored like the pileus. The entire plant becomes darker in drying, sometimes almost black. It is near *Russula nigricans*, but is smaller, and does not have a red juice as *R. nigricans* has.

CANTHARELLUS Adanson. [Pg 128]

From the other white-spored agarics of a fleshy consistency *Cantharellus* is distinguished by the form of the gills. The gills are generally forked, once or several times, in a dichotomous manner, though sometimes irregularly. They are blunt on the edge, not acute as in most of the other genera. The gills are usually narrow and in many species look like veins, folds, or wrinkles, but in some species, as in *Cantharellus aurantiacus*, they are rather thin and broad.

Figure 126.—Cantharellus cibarius. Under view showing forked gills with veins connecting them. Entire plant rich chrome yellow (natural size).

Cantharellus cibarius Fr. **Edible.**—This plant is known as the *chanterelle*. It has a very wide distribution and has long been regarded as one of the best of the edible mushrooms. Many of the writers on fungi speak of it in terms of high praise. The entire plant is a uniform rich chrome yellow. Sometimes it is symmetrical in form, but usually it is more or less irregular and unsymmetrical in form. The plants are 5–10 cm. high, the cap 4–8 cm. broad, and the stem short and rather thick.

Plate 41, Figure 127.—Cantharellus aurantiacus. Color orange yellow, and cap varies ochre, raw sienna, tawny, in different specimens (natural size). Copyright.

The **pileus** is fleshy, rather thick, the margin thick and blunt and at first inrolled. It is convex, becoming expanded or sometimes depressed by the margin of the cap becoming elevated. The margin is often wavy or repand, and in irregular forms it is only produced at one side, or more at one side than at the other, or the cap is irregularly [Pg 129] lobed. The **gills** are very narrow, stout, distant, more or less sinuous, forked or anastomosing irregularly, and because of the pileus being something like an inverted cone the gills appear to run down on the stem. The **spores** are faintly yellowish, elliptical, 7–10 μ. Figure 126 represents but a single specimen, and this one with a nearly lateral pileus.

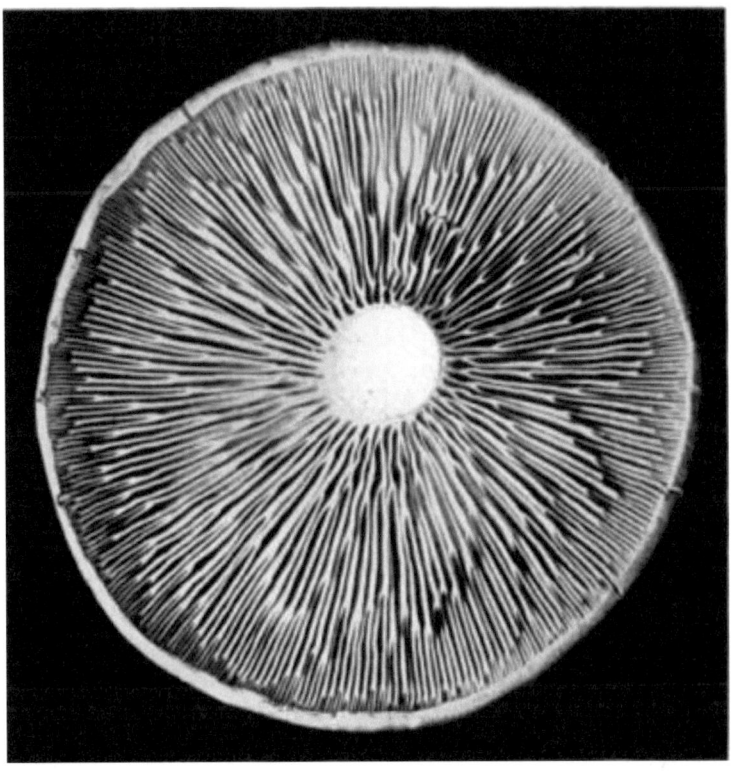

Figure 128.—Cantharellus aurantiacus, under view, enlarged nearly twice, showing regularly forked gills.

Cantharellus aurantiacus Fr.—This orange cantharellus is very common, and occurs on the ground or on very rotten wood, logs, branches, etc., from summer to very late autumn. It is widely distributed in Europe and America. It is easily known by its dull orange or brownish pileus, yellow gills, which are thin and regularly forked, [Pg 130] and by the pileus being more or less depressed or funnel-shaped. The plants are from 5–8 cm. high, the cap from 2–7 cm. broad, and the stem about 4–8 mm. in thickness.

The **pileus** is fleshy, soft, flexible, convex, to expanded, or obconic, plane or depressed, or funnel-shaped, the margin strongly inrolled when young, in age simply incurved, the margin plane or repand and undulate. The color varies from ochre yellow to dull orange, or orange ochraceous, raw sienna, and tawny, in different specimens. It is often brownish at the center. The surface of the pileus is minutely tomentose with silky hairs, especially toward the center, and sometimes smooth toward the margin. The flesh is 3–5 mm. at the center, and thin toward the margin. The gills are arcuate, decurrent, thin, the edge blunt, but not so much so as in a number of other species, crowded, regularly forked several times, at length ascending when the pileus is elevated at the margin. The color of the **gills** is orange to cadmium orange, or sometimes paler, cadmium yellow or deep chrome. The **stem** is clay color to ochre yellow, enlarged below, spongy, stuffed, fistulose, soft, fibrous, more or less ascending at the base.

The taste is somewhat nutty, sometimes bitterish. The plants in Fig. 127 (No. 3272, C. U. herbarium) were collected near Ithaca, October 7, 1899.

MARASMIUS Fr.

In this genus the plants are tough and fleshy or membranaceous, leathery and dry. They do not easily decay, but shrivel up in dry weather, and revive in wet weather, or when placed in water. This is an important character in distinguishing the genus. It is closely related to *Collybia*, from which it is difficult to separate certain species. On the other hand, it is closely related to *Lentinus* and *Panus*, both of which are tough and pliant. In *Marasmius*, however, the substance of the pileus is separate from that of the stem, while in *Lentinus* and *Panus* it is continuous, a character rather difficult for the beginner to understand. The species of *Marasmius*, however, are generally much smaller than those of *Lentinus* and *Panus*, especially those which grow on wood. The stem in *Marasmius* is in nearly all species central, while in *Lentinus* and *Panus* it is generally more or less eccentric. Many of the species of the genus *Marasmius* have an odor of garlic when fresh. Besides the fairy ring (*M. oreades*) which grows on the ground, *M. rotula* is a very common species on wood and leaves. It has a slender, black, shining stem, and a brownish pileus usually with a black spot in the depression in the [Pg 131] center. The species are very numerous. Peck, 23rd Report, N. Y.

State Mus., p. 124–126, describes 8 species. Morgan Jour. Cinn. Soc. Nat. Hist. **6**: 189–194, describes 17 species.

Marasmius oreades Fr. **Edible.**—This is the well known "fairy ring" mushroom. It grows during the summer and autumn in grassy places, as in lawns, by roadsides, in pastures, etc. It appears most abundantly during wet weather or following heavy rains. It is found usually in circles, or in the arc of a circle, though few scattered plants not arranged in this way often occur. The plants are 7–10 cm. high, the cap 2–4 cm. broad, and the stem 3–4 mm. in thickness.

Figure 129.—Marasmius oreades. Caps buff, tawny, or reddish.

The **pileus** is convex to expanded, sometimes the center elevated, fleshy, rather thin, tough, smooth, buff color, or tawny or reddish, in age, or in drying, paler. When moist the pileus may be striate on the margin. The **gills** are broad, free or adnexed, rounded near the stem, white or dull yellowish. The **spores** are elliptical, 7–8 μ long. The **stem** is tough, solid, whitish.

This widely distributed fungus is much prized everywhere by those who know it. It is not the only fungus which appears in rings, so that this habit is not peculiar to this plant. Several different kinds are known to appear in rings at times. The appearance of the fungus in rings is due to the mode of growth of the mycelium or spawn in the soil.

Having started at a given spot the mycelium consumes the food material in the soil suitable for it, and the plants for the first year appear in a group. In the center of this spot the mycelium, having consumed all the available food, probably dies after producing the [Pg 132] crop of mushrooms. But around the edge of the spot the mycelium or spawn still exists, and at the beginning of the next season it starts into growth and feeds on the available food in a zone surrounding the spot where it grew the previous year. This second year, then, the plants appear in a small ring. So in succeeding years it advances outward, the ring each year becoming larger. Where the plants appear only in the arc of a circle, something has happened to check or destroy the mycelium in the remaining arc of the circle.

It has been noted by several observers that the grass in the ring occupied by the mushrooms is often greener than that adjoining. This is perhaps due to some stimulus exerted by the mycelium of the fungus on the grass, or possibly the mycelium may in some way make certain foods available for the grass which gives an additional supply to it at this point.

Fig. 129 is from plants (No. 5503, C. U. herbarium) collected in a lawn, October 25, 1900, Ithaca.

Illustrations of some fine large rings formed by this fungus appeared in circular No. 13 by Mr. Coville, of the Division of Botany in the U. S. Dept. Agr.

Marasmius cohærens (Fr.) Bres. (*Mycena cohærens* Fr. *Collybia lachnophylla* Berk. *Collybia spinulifera* Pk.)—This plant grows in dense clusters, ten to twenty individuals with their stems closely joined below and fastened together by the abundant growth of threads from the lower ends. From this character the name *cohærens* was derived. The plants grow on the ground or on very rotten wood in woods during late spring and in the summer. The plant is not very common in this country, but appears to be widely distributed both in Europe and here, having been collected in Carolina, Ohio, Vermont, New York, etc. The plants are 12–20 cm. high, the cap 2–2.5 cm. broad, and the stem 4–7 mm. in thickness.

The **pileus** is fleshy, tough, convex or bell-shaped, then expanded, sometimes umbonate, or in age sometimes the margin upturned and more or less wavy, not viscid, but finely striate when damp, thin. The color varies from vinaceous cinnamon to chestnut or light leather color, or tawny, paler in age, and sometimes darker on the center. The **gills** are sometimes more or less crowded, narrow, 5–6 mm. broad, adnate, but notched, and sometimes becoming free from the stem. The color is light leather color, brick red or bay, the color and color variations being due to numbers of colored cystidia or spicules scattered over the surface of the gills and on the edge. The **cystidia** are fulvous, fusoid, 75–90 μ long. The **spores** are oval, white, small, 6 × 3 μ. The **stem** is long and slender, nearly cylindrical, tapering somewhat [Pg 133] above, slightly enlarged below, and rooting. The color is the same as that of the pileus or dark bay brown, and shining, and seems to be due to large numbers of spicules similar to those on the gills. The color is paler below in some cases, or gradually darker below in others. The stems are bound together below by numerous threads.

Figure 130 is from plants (No. 2373, C. U. herbarium) collected in woods near Freeville, N. Y. The plants have been collected near Ithaca on three different occasions, twice near Freeville about nine miles from Ithaca, and once in the woods at Ithaca. It is easily distinguished by its color and the presence of the peculiar setæ or cystidia.

Figure 130.—Marasmius cohaerens (Fr.) Bres. (= Mycena cohaerens Fr. = Collybia lachnophylla Berk. = C. spinulifera Pk.) Color
chestnut, light leather color, tawny or vinaceous cinnamon, darker
in center, stems dark, shining, gills leather color, or fulvous, or wine
color, brick red or bay, varying in different specimens (natural size).
Copyright.

Although the plant has been collected on several different occasions in America, it does not seem to have been recognized under

this name until recently, save the record of it from Carolina by de Schweinitz (Synop. fung. Car. No. 606. p. 81).

[Pg 134] LENTINUS Fr.

The plants of this genus are tough and pliant, becoming hard when old, unless very watery, and when dry. The genus differs from the other tough and pliant ones by the peculiarity of the gills, the gills being notched or serrate on the edges. Sometimes this appearance is intensified by the cracking of the gills in age or in drying. The nearest ally of the genus is *Panus*, which is only separated from *Lentinus* by the edge of the gills being plane. This does not seem a very good character on which to separate the species of the two genera, since it is often difficult to tell whether the gills are naturally serrate or whether they have become so by certain tensions which exist on the lamellæ during the expansion and drying of the pileus. Schrœter unites *Panus* with *Lentinus* (Cohn's Krypt. Flora, Schlesien, **3**, 1; 554, 1889). The plants are usually very irregular and many of them shelving, only a few grow upright and have regular caps.

Lentinus vulpinus Fr.—This is a large and handsome species, having a wide distribution in Europe and in this country, but it does not seem to be common. It grows on trunks, logs, stumps, etc., in the woods. It was quite abundant during late summer and in the autumn on fallen logs, in a woods near Ithaca. The **caps** are shelving, closely overlapping in shingled fashion (imbricated), and joined at the narrowed base. The surface is convex, and the margin is strongly incurved, so that each of the individual caps is shell-shaped (conchate). The surface of the pileus is coarsely hairy or hispid, the surface becoming more rough with age. Many coarse hairs unite to form coarse tufts which are stouter and nearly erect toward the base of the cap, and give the surface a tuberculate appearance. Toward the margin of the cap these coarse hairs are arranged in nearly parallel lines, making rows or ridges, which are very rough. The hairs and tubercles are dark in color, being nearly black toward the base, especially in old plants, and sometimes pale or of a smoky hue, especially in young plants. The pileus is flesh color when young, becoming darker when old, and the flesh is quite thin, whitish toward the gills and darker toward the surface. The

gills are broad, nearly white, flesh color near the base, coarsely serrate, becoming cracked in age and in drying, narrowed toward the base of the pileus, not forked, crowded, 4–6 mm. broad. The cap and gills are tough even when fresh. The plant has an intensely pungent taste.

Figures 131, 132 represent an upper, front, and under view of the pilei (No. 3315, C. U. herbarium).

[Pg 135]

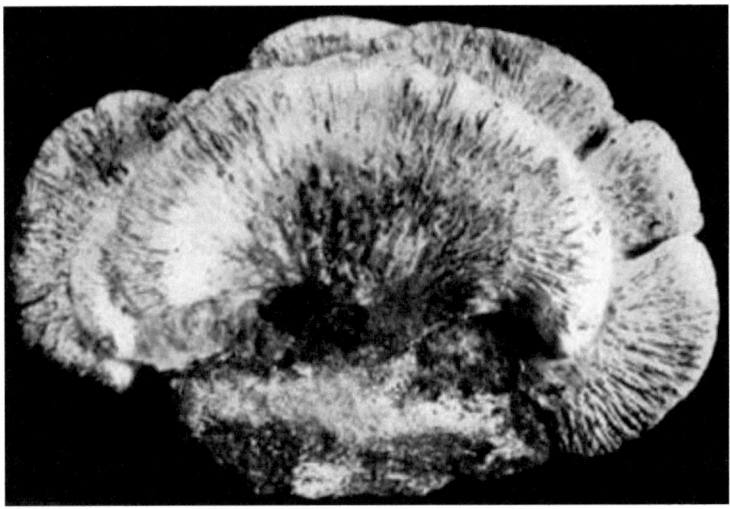

Plate 42, Figure 131.—Lentinus vulpinus. The coarse, hairy scales are black in old plants, paler, of a smoky hue, in younger ones (natural size). Copyright.

Lentinus lecomtei Fr., is a very common and widely distributed species growing on wood. When it grows on the upper side of logs the pileus is sometimes regular and funnel-shaped (cyathiform), but it is often irregular and produced on one side, especially if it grows on the side of the substratum. In most cases, however, there is a funnel-shaped depression above the attachment of the stem. The **pileus** is tough, reddish or reddish brown or leather color, hairy or sometimes strigose, the margin incurved. The **stem** is usually short, hairy, or in age it may become more or less smooth. The **gills** are narrow, crowded, the spores small, ovate to elliptical 5–6 × 2–3 μ.

According to Bresadola this is the same as *Panus rudis* Fr. It resembles very closely also *Panus cyathiformis* (Schaeff.) Fr., and *P. strigosus* B. & C.

Lentinus lepideus Fr., [*L. squamosus* (Schaeff.) Schroet.] is another common and widely distributed species. It is much larger than *L. lecomtei*, whitish with coarse brown scales on the cap. It is 12–20 cm. high, and the cap is often as broad. The stem is 2–8 cm. long and 1–2 cm. in thickness. It grows on wood.

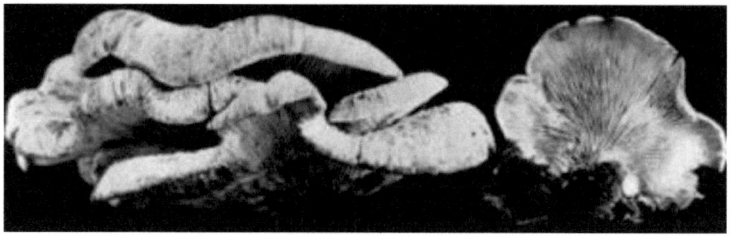

Figure 132.—Lentinus vulpinus, front and under view (natural size). Copyright.

Lentinus stipticus (Bull.) Schroet. (*Panus stipticus* Bull.) is a very small species compared with the three named [Pg 136] above. It is, however, a very common and widely distributed one, growing on wood, and may be found the year around. The pileus is 1–3 cm. in diameter, whitish or grayish, very tough, expanded in wet weather, and curled up in dry weather. The stem is very short, and attached to one side of the cap. When freshly developed the plant is phosphorescent.

SCHIZOPHYLLUM Fr.

This is a very interesting genus, but the species are very few. The plants are tough, pliant when fresh, and dry. The gills are very characteristic, being split along the edge and generally strongly revolute, that is, the split edges curve around against the side of the gill. This character can be seen sometimes with the aid of a hand lens, but is very evident when a section of the cap and gills is made and then examined with a microscope. The spores are white.

Figure 133.—Schizophyllum alneum (==S. commune). View of under side (natural size). Copyright.

Schizophyllum alneum (L.) Schroet.—This species usually goes by the name of *Schizophyllum commune*, but the earlier name is *S. alneum*. It is a very common plant and is world wide in its distribution, growing on wood, as on branches, trunks, etc. It is white, and the **pileus** is very hairy or tomentose, with coarse white hairs. It is 1–3 cm. in diameter, and the cap is sessile, either attached at one side when the cap is more produced [Pg 137] on one side than on the other, or it may be attached at or near the center of the top, when the cap is more evenly developed on all sides. It is often crenate or lobed on the margin, the larger plants showing this character more prominently. The margin is incurved. The **gills** are white, wooly, branched and extend out toward the margin of the cap like the radiations of a fan. The gills are deeply split along the edge, and strongly revolute. It is a very pretty plant, but one becomes rather tired of collecting it because it is so common. It may be found at all seasons of the year on dead sticks and branches, either in the woods or elsewhere, if the branches are present. It is very coriaceous, and tough. During dry weather it is much shrunken and curled up, but during rains it expands quickly and then it is seen in its beauty.

Plate 43, Figure 134. — Trogia crispa. Large cluster of caps, view of underside (natural size). Copyright.

Figure 133 shows the plant in the expanded condition, from the under side. The plants were growing on a hickory branch, and were dry and shrunken when brought in the laboratory. The branch and the fungus were placed in water for a few hours, when the fungus expanded, and was then photographed in this condition.

TROGIA Fr.

This genus is characterized, according to Fries, by the gills being channeled along the edge, but singularly the only species attributed to the genus in Europe and in our country has not channeled gills, but only somewhat crisped along the edges. It is usually, therefore, a difficult matter for a beginner to determine the plant simply from this description. The gills are furthermore narrow, irregular, and the plants are somewhat soft and flabby when wet, but brittle and persistent when dry, so that when moistened they revive and appear as if fresh.

Trogia crispa Fr. — This species is the principal if not only one in Europe and America. It is widely distributed, and sometimes not very uncommon. It occurs on trunks, branches, etc., often on the birch. The plants are from 0.5–1 cm. broad, usually sessile. The upper surface is whitish or reddish yellow toward the attachment, sometimes tan color, and when young it is sometimes covered with whitish hairs. The gills are very narrow, vein-like, irregular, interrupted or continuous, and often more or less branched. The gills are very much crisped, hence the name, blunt at the edge and white or bluish gray. The caps are usually much crowded and overlapped in an imbricated fashion as shown in Fig. 134; a photograph of a fine specimen after being moistened.

[Pg 138] CHAPTER VII.

THE ROSY-SPORED AGARICS.

The spores are rosy, pink, salmon colored, flesh colored, or reddish. For analytical keys to the genera see Chapter XXIV .

PLUTEUS Fr.

In the genus *Pluteus* the volva and annulus are both wanting, the gills are usually free from the stem, and the stem is easily broken out from the substance of the cap, reminding one in some cases of a ball and socket joint. The substance of the cap is thus said to be not continuous with that of the stem. The spores seen in mass are flesh colored as in other genera of this subdivision of the agarics.

Figure 135.—Pluteus cervinus. Cap grayish brown, or sooty, smooth or sometimes scaly, rarely white, stem same color, but paler; gills first white, then flesh color (natural size, often larger). Copyright.

Pluteus cervinus Schaeff. **Edible.**—This is one of the very common species of the higher fungi, and is also very widely distributed. It [Pg 139] varies considerably in size and appearance. It is 7–15 cm. high, the cap 5–10 cm. broad, and the stem 6–12 mm. in thickness. It occurs on the ground from underground roots or rotten wood, or grows on decaying stumps, logs, etc., from spring until late autumn. Sometimes it is found growing in sawdust.

The **pileus** is fleshy, bell-shaped, then convex, and becoming expanded, the surface usually smooth, but showing radiating fibrils, grayish brown, or sometimes sooty, sometimes more or less scaly. The **gills** are not crowded, broad, free from the stem, white, then becoming flesh color with the maturity of the spores. One very characteristic feature of the plant is the presence of **cystidia** in the hymenium on the gills. These are stout, colorless, elliptical, thick-walled, and terminate in two or three blunt, short prongs.

The **stem** is nearly equal, solid, the color much the same as that of the pileus, but often paler above, smooth or sometimes scaly.

In some forms the plant is entirely white, except the gills. In addition to the white forms occurring in the woods, I have found them in an old abandoned cement mine growing on wood props.

[Pg 140]

Figure 136.—Pluteus tomentosulus. Cap and stem entirely white, gills flesh color, stem furrowed and tomentose (natural size). Copyright.

Pluteus tomentosulus Pk.—This plant was described by Peck in the 32d Report, N. Y. State Mus., page 28, 1879. It grows on decaying wood in the woods during July and August. The plants are 5–12 cm. high, the cap 3–7 cm. broad, and the stem 4–8 mm. in thickness. The description given by Peck is as follows: "Pileus thin, convex or expanded, subumbonate, dry, minutely squamulose-tomentose, white, sometimes pinkish on the margin; lamellæ rather broad, rounded behind, free, crowded, white then flesh colored; stem equal, solid, striate, slightly pubescent or subtomentose, white; spores subglobose, 7 μ in diameter, generally containing a large single nucleus." From the plant collected at Ithaca the following notes were made. The **pileus** and stem are entirely white, the gills flesh color. The pileus is expanded, umbonate, thin except at the umbo, minutely floccose squamulose, no pinkish tinge noted; the flesh is white, but on the umbo changing to flesh color where wounded. The **gills** are free, with a clear white space between stem and rounded edges, crowded, narrow (about 3–4 mm. broad) edge finely fimbriate, probably formed by numerous bottle-shaped cystidia on the edge, and which extend up a little distance on the side of the gills, but are not distributed in numbers over the surface of the gills; **cystidia** thin walled, hyaline. The **spores** are flesh colored, subglobose, 5–7 μ. **Stem** cylindrical, even, twisted somewhat, white, striate and minutely squamulose like the pileus, but with coarser scales, especially toward the base, solid, flesh white.

The species received its name from the tomentose, striate character of the stem. The plants (No. 3219, C. U. herbarium) illustrated in Fig. 136 were collected in Enfield Gorge, vicinity of Ithaca, July 28, 1899.

VOLVARIA Fr.

This genus takes its name from the volva, which means a wrapper, and which, as we know from our studies of *Amanita*, entirely envelops the plant at a young stage. The genus is characterized then by the rosy or reddish spores, the presence of a volva, and the annulus is wanting. The stem is easily separable from the pileus at its junction, in this respect being similar to *Amanita*, *Amanitopsis*, *Lepiota* and others. The gills are usually, also, free from the stem. The species grow on rotting wood, on leaf mould and on richly manured ground, etc. They are of a very soft texture and usually soon decay.

Volvaria bombycina (Pers.) Fr. **Edible.**—The silky volvaria is so called because of the beautiful silky texture of the surface of the cap. It is not very common, but is world wide in its distribution, and occurs [Pg 141] on decayed wood of logs, stumps, etc., during late summer and in autumn. It is usually of a beautiful white color, large, the volva large and thick, reminding one of a bag, and the stem is ascending when the plant grows on the side of the trunk, or erect when it grows on the upper side of a log or stump. The plant is from 8–16 cm. high, the cap 6–20 cm. broad, and the stem 1–1.5 cm. thickness.

The **pileus** is globose, then bell-shaped, and finally convex and somewhat umbonate, white, according to some becoming some-what reddish. The entire surface is silky, and numerous hairs stand out in the form of soft down, when older the surface becoming more or less scaly, or rarely becoming smooth at the apex. The flesh is white. The **gills** are crowded, very broad along the middle, flesh colored, the edge sometimes ragged. The **spores** are rosy in mass, oval to broadly elliptical, 6–9 × 5–6 μ, smooth. The **stem** tapers from the base to the apex, is solid, smooth. The **volva** is large and bag-like. The plant is considered edible by some. Figure 137 is from a plant (No. 3096, C. U. herbarium) collected on a log of Acer rubrum in Cascadilla woods, Ithaca, on August 10th, 1898.

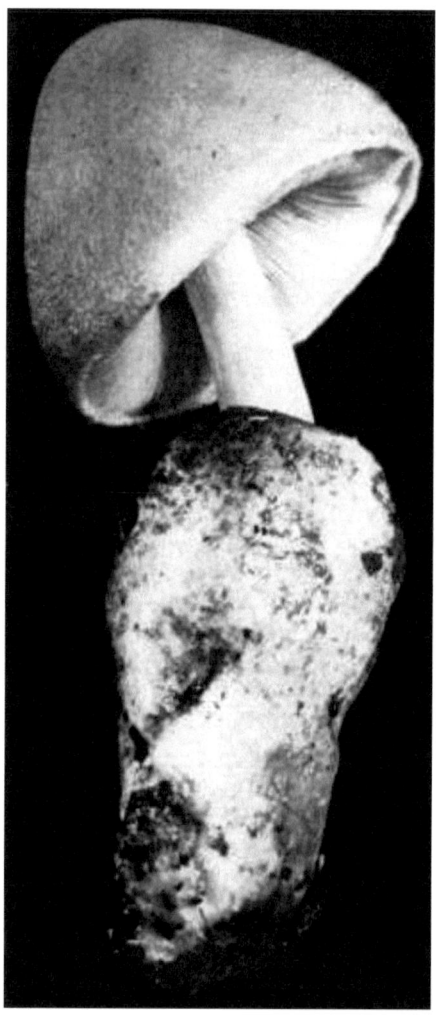

Figure 137.—Volvaria bombycina. Cap, stem and volva entirely white, gills flesh color (natural size). Copyright.

Volvaria speciosa Fr.—This plant seems to be rare, but it has a wide distribution in Europe and the United States. It occurs on richly manured ground, on dung, etc. The plants are 10–20 cm. high, the cap 6–12 cm. broad, and the stem 1–2 cm. in thickness. The entire plant is white or whitish, sometimes grayish, especially at the center, where it is also sometimes darker and of a smoky color. [Pg 142]

The **pileus** is globose when young, then bell-shaped, and finally more or less expanded, and umbonate, smooth, very viscid, so that earth, leaves, etc., cling to it. The flesh is white and very soft. The **gills** are free, flesh colored to reddish or fulvous, from the deeply colored spores. The **spores** are broadly elliptical, or oval, 12–18 × 8–10 μ. The **stem** is nearly cylindrical, or tapering evenly from the base, when young more or less hairy, becoming smooth. The **volva** is large, edge free, but fitting very close, flabby and irregularly torn.

The species is reported from California by McClatchie, and from Wisconsin by Bundy.

Specimens were received in June, 1898, from Dr. Post of Lansing, Mich., which were collected there in a potato patch. It was abundant during May and June. Plants which were sent in a fresh condition were badly decayed by the time they reached Ithaca, and the odor was very disagreeable. It is remarkable that the odor was that of rotting potatoes! In this connection might be mentioned Dr. Peck's observation (Bull. Torr. Bot. Club 26: p. 67, 1899) that *Agaricus maritimus* Pk., which grows near the seashore, possessed "a taste and odor suggestive of the sea."

McClatchie reports that it is common in cultivated soil, especially grain fields and along roads, and that it is "a fine edible agaric and our most abundant one in California."

CLITOPILUS Fr.

In the rosy-spored agarics belonging to this genus the gills are decurrent, that is, extend for some distance down on the stem. The stem is fleshy. The gills are white at first and become pink or salmon color as the plants mature, and the spores take on their characteristic color. The plants should thus not be confused with any of the species of *Agaricus* to which the common mushroom belongs, since in those species the gills become dark brown or blackish when mature. The genus corresponds with *Clitocybe* among the white-spored ones.

Clitopilus prunulus Scop. **Edible.**—This species grows on the ground in the woods from mid-summer to autumn. It is not very common, but sometimes appears in considerable quantities at one place. During the autumn of 1898 quite a large number of specimens were found in a woods near Ithaca, growing on the ground around an old stump. The plants are 3–8 cm. high, the cap 5–10 cm. broad, and stem 1–2 cm. in thickness.

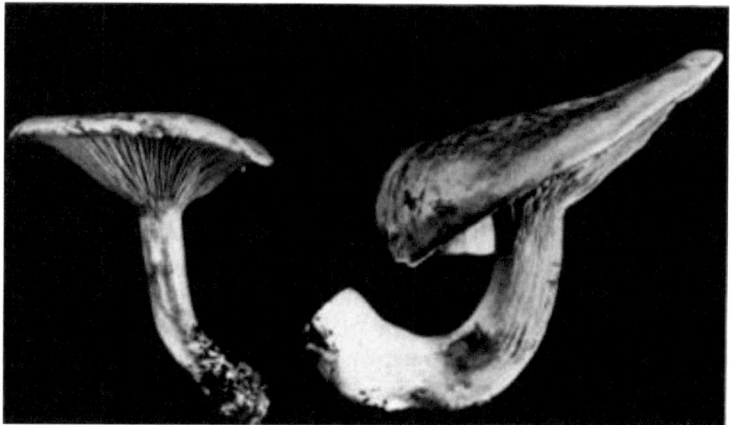

Plate 44, Figure 138.—Clitopilus prunulus, cap whitish or dark gray, gills flesh color (natural size). Copyright.

The **pileus** is fleshy, firm, convex and becoming nearly plane, and sometimes as the plants become old the center may be slightly depressed. [Pg 143] It is whitish in color, or dark gray, or with a leaden tint, dry, sometimes with a distinct bloom on the surface, and the margin is often wavy. The cap is sometimes produced more on one side than on the other. The **gills** are not close, at first whitish, then salmon colored as the spores mature, and they are decurrent as is characteristic of the genus. The **spores** are elliptical or nearly so, and measure 10–12 µ long.

Figure 138 is from plants collected near Ithaca, in the autumn of 1898. This species is considered to be one of the excellent mushrooms for food. When fresh it has a mealy odor and taste, as do several of the species of this genus. It is known as the prune mushroom.

Clitopilus orcella Bull. **Edible.**—This plant is sometimes spoken of as the sweet-bread mushroom. It is much like the prune mushroom just described, in odor and taste, and sometimes resembles it in form and other characters. It is white in color, and the plants are usually considerably smaller, and the pileus is, according to my observations, sometimes more irregular, lobed and wavy on the margin. The flesh is also softer, and the cap is said to be slightly viscid in wet weather. The plant grows in the woods and sometimes in open fields.

ENTOLOMA Fr.

The volva and annulus are absent in this genus, the spores are rosy, the gills adnate to sinuate or adnexed, easily separating from the stem in some species. The stem is fleshy or fibrous, sometimes waxy, and the pileus is fleshy with the margin incurved, especially when young. The spores are prominently angular. The genus corresponds with *Tricholoma* of the white-spored agarics, and also with *Hebeloma* and *Inocybe* of the ochre-spored ones. *Entoloma repandum* Bull., is an *Inocybe* [*I. repandum* (Bull.) Bres.] and has angular spores resembling those of an *Entoloma*, but the spores are not rosy.

Entoloma jubatum Fr.—Growing on the ground in woods. The plants are 5–10 cm. high, the cap 3–6 cm. broad, and the stem 3–6 mm. in thickness.

The **pileus** is conic in some plants, to convex and umbonate, thin, minutely scaly with blackish hairy scales, dull heliotrope purple, darker on the umbo. The **gills** are vinaceous rufus to deep flesh color, strongly sinuate, and irregularly notched along the edge. The **spores** are irregularly oval to short oblong, coarsely angular, with an oil drop, 5–7 angled, 7–11 × 6–7 μ. The **stem** is of the same color as the pileus, sometimes deeply rooting, hollow. Figure 139 is from [Pg 144] plants (No. 4000, C. U. herbarium) collected at Blowing Rock, N. C., during September, 1899.

Entoloma grayanum Pk.—This plant grows on the ground in woods. It is from 6–8 cm. high, the cap is 3–6 cm. broad, and the stem 4–6 mm. in thickness.

Figure 139. — Entoloma jubatum. Entire plant dull heliotrope purple, gills later flesh color (natural size). Copyright.

The **pileus** is convex to expanded, sometimes broadly umbonate, drab in color, the surface wrinkled or rugose, and watery in appearance. The flesh is thin and the margin incurved. The **gills** are first drab in color, but lighter than the pileus, becoming pinkish in age. [Pg 145] The **spores** on paper are very light salmon color. They are globose or rounded in outline, 5–7 angled, with an oil globule, 8–10 μ in diameter. The **stem** is the same color as the pileus, but lighter, striate, hollow, somewhat twisted, and enlarged below. Figure 140 is from plants (No. 3998, C. U. herbarium) collected at Blowing Rock, N. C., during September, 1899.

Figure 140. — Entoloma grayanum. Cap and stem drab, gills flesh color (natural size). Copyright.

Entoloma strictius Pk. — The plants grow in grassy places, pastures, etc. They are clustered, sometimes two or three joined at the base of the stem. They are 7–10 cm. high, the caps 2–4 cm. broad, and the stems 3–6 mm. in thickness.

The **pileus** is convex, the disk expanded, and the margin incurved and more or less wavy or repand on the extreme edge. It is umbonate at the center with usually a slight depression around the umbo, smooth, watery (hygrophanous) in appearance, not viscid, of an umber color, shining, faintly and closely striate on the margin. [Pg 146] In drying the surface of the pileus loses some of its dark

237

umber color and presents a silvery sheen. The flesh is fibrous and umber color also. The **gills** are grayish white, then tinged with flesh color, slightly sinuate, the longer ones somewhat broader in the middle (ventricose), rather distant, and quite thick as seen in cross section, the center of the gill (trama) presenting parallel threads. The sub-hymenium is very thin and composed of small cells; the **basidia** are clavate, 25–30 × 9–10 μ, and four-spored. The **spores** are dull rose color on paper, subgloblose, 5–8 μ in diameter, angular with 5–6 angles as seen from one side. The **stem** is the same color as the pileus, but considerably lighter. It is hollow with white fibers within, fibrous striate on the surface, twisted, brittle, and somewhat cartilaginous, partly snapping, but holding by fibers in places, cylindrical, even, ascending, with delicate white fibers covering the lower end.

Figure 141.—Entoloma strictius. Cap umber or smoky, stem paler, gills grayish, then flesh color (natural size). Copyright.

Figure 141 is from plants (No. 2461, C. U. herbarium) collected near Ithaca, October, 1898.

LEPTONIA Fr.

In *Leptonia* the stem is cartilaginous, hollow or stuffed, smooth and somewhat shining. The pileus is thin, umbilicate or with the center darker, the surface hairy or scaly, and the margin at first incurved. The gills are adnate or adnexed at first, and easily separating from the stem in age. Many of the species are bright colored.

Figure 142.—Leptonia asprella. Cap hair brown (mouse colored), minute dark scales at center, stem same color, but sometimes reddish brown, green or blue, gills flesh color.

Leptonia asprella Fr.—This species occurs on the ground in woods or in open grassy places. The plants are 3–5 cm. high, the cap 2–4 cm. broad, and the stem 2–3 mm. in thickness.

The **pileus** is convex, then more or less expanded, umbilicate, rarely umbonate, hair brown (mouse colored), with dark scales on the center and minute scales over the surface, striate.

The **gills** are sinuate to adnexed. The **spores** are strongly 5–6 angled, 10–12 × 8–10 μ. The **stem** is smooth, even, usually the same color as the cap, but sometimes it is reddish brown, green, or blue. Figure 142 is from plants (No. 3996, C. U. herbarium) collected at Blowing Rock, N. C., during September, 1899.

Leptonia incana Fr., is a more common species, and is characterized by an odor of mice.

[Pg 148] ECCILIA Fr.

The genus *Eccilia* corresponds with *Omphalia* of the white-spored agarics. The stem is cartilaginous, hollow or stuffed. The pileus is thin and somewhat membranaceous, plane or depressed at the center, and the margin at first incurved. The gills are more or less decurrent.

Eccilia polita Pers. — This plant occurs on the ground in woods. It is 6–10 cm. high, the cap 2–4 cm. broad, and the stem is 3–4 mm. in thickness.

Figure 143.—Eccilia polita. Cap hair brown to olive, stem lighter, gills flesh color, notched and irregular (natural size). Copyright.

The **pileus** is convex and <u>umbilicate</u>, somewhat membranaceous, smooth, watery in appearance, finely striate on the margin, hair brown to olive in color. The **gills** are decurrent. In the specimens illustrated in Fig. 143 the gills are very irregular and many of them appear sinuate. The **spores** are strongly 4–5 angled, some of them square, 10–12 µ in diameter, with a prominent mucro at one angle. The **stem** is cartilaginous, becoming hollow, lighter in color than the pileus, and somewhat enlarged below. Figure 143 is from plants (No. 3999, C. U. herbarium) collected at Blowing Rock, N. C., during September, 1899.

Plate 45, Figure 144.—Claudopus nidulans, view of under side. Cap rich yellow or buff, gills flesh color (natural size). Copyright.

CLAUDOPUS W. Smith. [Pg 149]

In the genus *Claudopus*, recognized by some, the pileus is eccentric or lateral, that is, the stem is attached near the side of the cap, or the cap is sessile and attached by one side to the wood on which the plant is growing; or the plants are resupinate, that is, they may be spread over the surface of the wood.

The genus is perhaps not well separated from some of the species of *Pleurotus* with lilac spores like *P. sapidus*. In fact, a number of the species were formerly placed in *Pleurotus*, while others were placed in *Crepidotus* among the ochre-spored agarics. Several species are reported from America. Peck in 39th Report N. Y. State Mus., p. 67, *et seq.*, 1886, describes five species.

Claudopus nidulans (Pers.) Pk.—This is one of the very pretty agarics growing on dead branches and trunks during the autumn, and is widely distributed. It has, however, been placed in the genus *Pleurotus*, as *P. nidulans*. But because of the pink color of the spores in mass, Peck places it in the genus *Claudopus*, where Fries suggested it should go if removed from *Pleurotus*. It seems to be identical with *Panus dorsalis Bosc*. It is usually sessile and attached to the side of dead branches, logs, etc., in a shelving manner, or sometimes it is resupinate.

The **pileus** is sessile, or sometimes narrowed at the base into a short stem, the caps often numerous and crowded together in an overlapping or imbricate manner. It is nearly orbicular, or reniform, and 1–5 cm. broad. The margin is at first involute. The surface is coarsely hairy or tomentose, or scaly toward the margin, of a rich yellow or buff color. It is soft, but rather tough in consistency. The **gills** are broad, orange yellow. The **spores**, pink in mass, are smooth, elongated, somewhat curved, 6–8 µ long.

Figure 144 is from plants (No. 2660, C. U. herbarium) collected in woods near Ithaca.

[Pg 150] CHAPTER VIII.

THE OCHRE-SPORED AGARICS.

The spores are ochre yellow, rusty, rusty-brown, or some shade of yellow. For analytical keys to the genera see Chapter XXIV .

PHOLIOTA Fr.

The genus *Pholiota* has ferruginous or ferruginous brown spores. It lacks a volva, but has an annulus; the gills are attached to the stem. It then corresponds to *Armillaria* among white-spored agarics, and *Stropharia* among the purple-brown-spored ones. There is one genus in the ochre or yellow-spored plants with which it is liable to be confused on account of the veil, namely *Cortinarius*, but in the latter the veil is in the form of loose threads, and is called an arachnoid veil, that is, the veil is spider-web-like. Many of the species of *Pholiota* grow on trunks, stumps, and branches of trees, some grow on the ground.

Pholiota præcox Pers. **Edible.**—*Agaricus candicans* Bull. T. 217, 1770: *Pholiota candicans* Schroeter, Krypt, Flora, Schlesien, p. 608, 1889. This plant occurs during late spring and in the summer, in pastures, lawns and grassy places, roadsides, open woods, etc. Sometimes it is very common, especially during or after prolonged or heavy rains. The plants are 6–10 cm. high, the cap from 5–8 cm. broad, and the stem 3–5 mm. in thickness. The plants are scattered or a few sometimes clustered.

The **pileus** is convex, then expanded, whitish to cream color or yellowish, then leather color, fleshy, the margin at first incurved, moist, not viscid. Sometimes the pileus is umbonate. The surface is sometimes uneven from numerous crowded shallow pits, giving it a frothy appearance. In age the margin often becomes upturned and fluted. The **gills** are adnate or slightly decurrent by a tooth, 3–4 mm. broad, a little broader at or near the middle, crowded, white, then ferruginous brown, edge sometimes whitish. There is often a prominent angle in the gills at their broadest diameter, not far from the stem, which gives to them, when the plants are young or middle age, a sinuate appearance. The **spores** are ferruginous brown, elliptical. **Cystidia** abruptly club-shaped, with a broad apiculus. The [Pg 151] **stem** is stuffed, later fistulose, even, fragile, striate often above the annulus. The stem is whitish or sometimes flesh color. The veil is whitish, large, frail, and sometimes breaks away from the stem and clings in shreds to the margin of the cap.

Plate 46, Figure 145.—Pholiota praecox. Cap whitish, to cream, or leather color, stem white, gills white then ferruginous brown (natural size). Copyright.

Figure 145 is from plants (No. 2362, C. U. herbarium) collected on the campus of Cornell University, June, 1898. The taste is often slightly bitter.

Pholiota marginata Batsch.—This is one of the very common species, a small one, occurring all during the autumn, on decaying trunks, etc., in the woods. The plants are usually clustered, though appearing also singly. They are from 4–10 cm. high, the cap 3–4 cm. broad, and the stem 3–5 μ in thickness.

Plate 47, Figure 146. — Pholiota adiposa. Cap very viscid, saffron-yellow or burnt umber or wood-brown in center, scales wood-brown to nearly black, stem whitish then yellowish; gills brownish, edge yellow (natural size, sometimes larger). Copyright.

The **pileus** is convex, then plane, tan or leather colored, darker when dry. It has a watery appearance (hygrophanous), somewhat fleshy, smooth, striate on the margin. The **gills** are joined squarely to the stem, crowded, at maturity dark reddish brown from the spores.

Figure 147.—Pholiota marginata. Cap and stem tan or leather color, gills dark reddish brown when mature (natural size). Copyright.

The **stem** is cylindrical, equal, smooth, fistulose, of the same color as the pileus, becoming darker, and often with whitish fibrils at the base. The **annulus** is distant from the apex of the stem, and often disappears soon after the expansion of the pileus. Figure 147 is from plants (No. 2743, C. U. herbarium) collected near Ithaca.

Pholiota unicolor Vahl, is a smaller plant which grows in similar situations. The plants are usually clustered, 3–5 cm. high, and the caps 6–12 mm. in diameter, the annulus is thin but entire and persistent. The entire plant is bay brown, becoming ochraceous in color, and the margin of the cap in age is striate, first bell-shaped, then convex and somewhat umbonate. The gills are lightly adnexed.

Pholiota adiposa Fr.—The fatty pholiota usually forms large clusters during the autumn, on the trunks of trees, stumps, etc. It is sometimes of large size, measuring up to 15 cm. and the pileus up to [Pg 152] 17 cm. broad. Specimens collected at Ithaca during October, 1899, were 8–10 cm. high, the pileus 4–8 cm. broad, and the stems 5–9 mm. in thickness. The plants grew eight to ten in a cluster and the bases of the stems were closely crowded and loosely joined.

The **pileus** is convex, then expanded, the margin more or less inrolled, then incurved, prominently umbonate, very viscid when moist, the ground color a saffron yellow or in the center burnt umber to wood brown. The cuticle of the pileus is plain or torn into scales which are wood brown, or when close together they are often darker, sometimes nearly black. The flesh is saffron yellow, thick at the center of the cap, thinning out toward the margin, spongy and almost tasteless. The **gills** are adnate, and sometimes a little notched, brown (mars brown), and the edge yellow, 6–7 mm. broad. The **spores** are 8 × 5 μ. The **stem** tapers downward, is compact, whitish then yellow, saffron yellow, flesh vinaceous, viscid, and clothed more or less with reflexed (pointing downward) scales. The stem is somewhat cartilaginous, tough, but snapping off in places. The veil is thin floccose and sometimes with coarse scales, soon disappearing.

Figure 146 is from plants (No. 3295, C. U. herbarium) collected on the Ithaca flats from a willow trunk, Oct. 10, 1899.

Pholiota aurivella Batsch, which has been found in the United States, is closely related to *P. adiposa*.

Pholiota squarrosa Müll., widely distributed and common in the autumn, both in Europe and America, on stumps and trunks, is a large, clustered, scaly plant, the scales "squarrose", and abundant over the pileus and on the stem below the annulus. It is brownish or ferruginous in color.

Pholiota squarrosoides Pk., as its name indicates, is closely related to *P. squarrosa*. It has erect, pointed, persistent scales, especially when young, and has a similar habit to *squarrosa*, but differs chiefly in the pileus being viscid, while that of *P. squarrosa* is dry. *P. subsquarrosa* Fr., occurring in Europe, and also closely related to *P. squarrosa*, is viscid, the scales are closely appressed to the surface of the cap, while in *squarrosa* they are prominent and revolute.

Pholiota cerasina Pk., occurs on decaying trunks of trees during late summer. The plants grow in tufts. They are 5–12 cm. high, the caps 5–10 cm. in diameter, and the stems 4–8 mm. in thickness. The pileus is smooth, watery when damp, cinnamon in color when fresh, becoming yellowish in drying, and the flesh is yellowish. The stem is solid, and equal, the apex mealy. The annulus is not persistent, and the gills are crowded and notched. The spores are elliptical, and rugose, 5 × 8 μ.

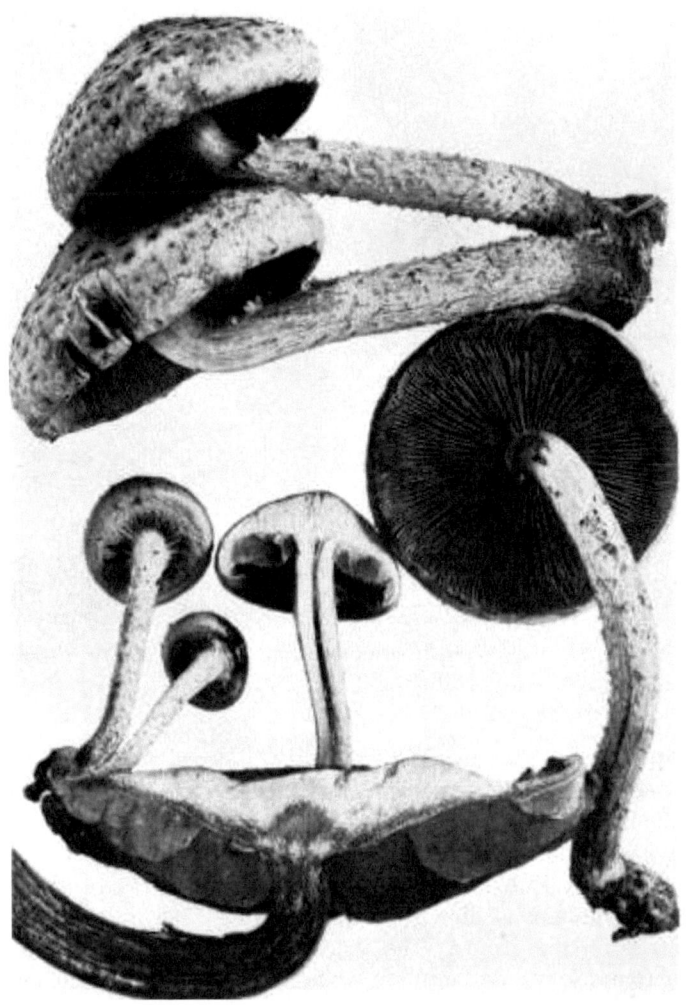

Plate 48, Figure 148.—Pholiota squarrosoides. Entire plant brownish or reddish brown; pileus viscid (three-fourths natural size). Copyright.

[Pg 153]

Plate 49, Figure 149.—Pholiota johnsoniana. Cap yellowish to yellowish brown, stem whitish, gills grayish then rust-brown (natural size). Copyright.

Pholiota johnsoniana Pk. **Edible.**—This species was described from specimens collected at Knowersville, N. Y., in 1889, by Peck, in the 23rd Report N. Y. State Mus., p. 98, as *Agaricus johnsonianus*. I found it at Ithaca, N. Y., for the first time during the summer of 1899, and it was rather common during September, 1899, in the Blue Ridge Mountains at Blowing Rock, N. C. It grows in woods or in pastures on the ground. The larger and handsomer specimens I have found in rather damp but well drained woods. The plants are 7–15 cm. high, the cap 5–10 cm. broad, and the stem 6–12 mm. in thickness.

The **pileus** is fleshy, very thick at the center, convex, then expanded and plane, smooth, sometimes finely striate on the thin margin when moist, yellowish, or fulvous, the margin whitish. The **gills** are attached to the stem by the upper angle (adnexed), rounded, or some of them angled, some nearly free. In color they are first gray, then rusty brown. They appear ascending because of the somewhat top-shaped pileus. The **spores** are irregularly ovoid, 4–6 × 3–3.5 μ. The **stem** is cylindrical or slightly tapering upward, smooth, slightly striate above the annulus, whitish, solid, with a tendency to become hollow. The **veil** is thick, and the annulus narrow and very thick or "tumid," easily breaking up and disappearing. The plant is quite readily distinguished by the form of the pileus with the ascending gills and the tumid annulus. Peck says it has a "somewhat nutty flavor."

Figure 149 is from plants (No. 4014, C. U. herbarium) collected at Blowing Rock, N. C., during September, 1899.

NAUCORIA Fr.

This genus, with ferruginous spores, corresponds with *Collybia* among the white-spored agarics. The gills are free or attached, but not decurrent, and the stem is cartilaginous. The plants grow both on the ground and on wood. Peck, 23rd Report N. Y. State Mus., p. 91, *et seq.*, gives a synopsis of seven species.

Naucoria semi-orbicularis Bull. **Edible.**—This is one of the common and widely distributed species. It occurs in lawns, pastures, roadsides, etc., in waste places, from June to autumn, being more abundant in rainy weather. The plants are 7–10 cm. high, the cap 3–5 cm. broad, and the stem 2–3 mm. in thickness. The **pileus** is convex to expanded, and is remarkably hemispherical, from which the species takes the name of *semi-orbicularis*. It is smooth, viscid when moist, tawny, and in age ochraceous, sometimes the surface is cracked into areas. The **gills** are attached, sometimes notched, crowded, [Pg 154] much broader than the thickness of the pileus, pale, then reddish brown. The **stem** is tough, slender, smooth, even, pale reddish brown, shining, stuffed with a whitish pith. Peck says that the plants have an oily flavor resembling beechnuts.

Naucoria vernalis Pk.—*Naucoria vernalis* was described by Peck in 23rd Report N. Y. State Mus., p. 91, from plants collected in May. The plants described here appeared in woods in late autumn. The specimens from which this description is drawn were found growing from the under side of a very rotten beech log, usually from deep crevices in the log, so that only the pileus is visible or exposed well to the view. The plants are 4–8 cm. high, the cap 2–3 cm. broad, and the stem 4–5 mm. in thickness. The taste is bitter.

Figure 150.—Naucoria vernalis. Cap hair brown to clay color; gills grayish brown to wood brown; stem clay color (natural size). Copyright.

The **pileus** is convex, then the center is nearly or quite expanded, the margin at first inrolled and never fully expanded, hygrophanous, smooth (not striate nor rugose), flesh about 5–6 mm. thick at center, thin toward the margin. The color changes during growth, it is from ochraceous rufus when young (1–2 mm. broad), then clove brown to hair brown and clay color in age. The **gills** are grayish brown to wood brown, at first adnate to slightly sinuate, then easily breaking away and appearing adnexed. The **spores** are wood brown in color, oval to short elliptical and inequilateral 6–8 × 4–5 μ. **Cystidia** hyaline, bottle shaped, 40–50 × 8–12 μ. The **stem** is somewhat hollow and stuffed, rather cartilaginous, though somewhat brittle, especially [Pg 155] when very damp, breaking out from the pileus easily though with fragments of the gills remaining attached, not strongly continuous with the substance of the pileus. The color is buff to pale clay color; the stem being even, not bulbous but somewhat enlarged below, mealy over the entire length, which may be washed off by rains, striate at apex either from marks left by the

252

gills or remnants of the gills as they become freed from the stem. Base of stem sometimes with white cottony threads, especially in damp situations. In the original description the stem is said to be "striate sulcate." Figure 150 is from plants (No. 3242, C. U. herbarium) collected in woods near Ithaca, October 1, 1899.

GALERA Fr.

Galera with ochraceous (ochraceous ferruginous) spores corresponds to *Mycena* among the white-spored agarics. The pileus is usually bell-shaped, and when young the margin fits straight against the stem. The stem is somewhat cartilaginous, but often very fragile. The genus does not contain many species. Peck gives a synopsis of five American species in the 23rd Report N. Y. State Mus., p. 93, *et seq.*, and of twelve species in the 46th Report, p. 61, *et seq.* One of the common species is **Galera tenera** Schaeff. It occurs in grassy fields or in manured places. The plants are 5–8 cm. high, the cap 8–16 mm. broad, and the stem 2–3 mm. in thickness. The **pileus** is oval to bell-shaped, and tawny in color, thin, smooth, finely striate, becoming paler when dry. The **gills** are crowded, reddish-brown, adnexed and easily separating. The **stem** is smooth, colored like the pileus but a little paler, sometimes striate, and with mealy whitish particles above. **Galera lateritia** is a related species, somewhat larger, and growing on dung heaps and in fields and lawns. **Galera ovalis** Fr., is also a larger plant, somewhat shorter than the latter, and with a prominent ovate cap when young. **Galera antipoda** Lasch., similar in general appearance to G. *tenera*, has a rooting base by which it is easily known. **Galera flava** Pk., occurs among vegetable mold in woods. The pileus is membraneous, ovate or campanulate, moist or somewhat watery, obtuse, plicate, striate on the margin, yellow. The plants are 5–8 cm. high, the caps 12–25 mm. broad, and the stem 2–3 mm. in thickness. The plant is recognized by the pale yellow color of the caps and the plicate striate character of the margin. The plicate striate character of the cap is singular among the species of this genus, and is shared by another species, **G. coprinoides** Pk.

[Pg 156] FLAMMULA Fr.

In the genus *Flammula*, the pileus is fleshy, stem fleshy-fibrous, and the gills adnate to decurrent.

Figure 151.—Flammula polychroa, under view. Cap vinaceous buff to orange buff, scales lilac, purple or lavender; gills drab to hair brown (natural size). Copyright.

Flammula polychroa Berk.—This is a beautiful plant with tints of violet, lavender, lilac and purple, especially on the scales of the pileus, on the veil and on the stem. It occurs in clusters during late summer and autumn, on logs, branches, etc., in the woods. The plants occur singly, but more often in clusters of three to eight or more. The plants are 4–7 cm. high, the cap 3–5 cm. broad, and the stem 4–6 mm. in thickness.

The **pileus** is convex, and in the young stage the margin strongly incurved, later the cap becomes expanded and has a very broad umbo. It is very viscid. The surface is covered with delicate hairs which form scales, more prominent during mid-age of the plant, and on the margin of the cap. These scales are very delicate and vary in color from vinaceous-buff, lilac, wine-purple, or lavender.

The ground color of the pileus is vinaceous-buff or orange-buff, and toward the margin often with shades of beryl-green, especially where it has been [Pg 157] touched. In the young plants the color of the delicate hairy surface is deeper, often phlox-purple, the color becoming thinner as the cap expands.

The **gills** are notched (sinuate) at the stem, or adnate, sometimes slightly decurrent, crowded. Before exposure by the rupture of the veil they are cream-buff in color, then taking on darker shades, drab to hair brown or sepia with a purple tinge. The **stem** is yellowish, nearly or quite the color of the cap, often with a purplish tinge at the base. It is covered with numerous small punctate scales of the same color, or sulphur yellow above where they are more crowded and larger. The scales do not extend on the stem above the point where the veil is attached. The stem is slightly striate above the attachment of the veil. It is somewhat tough and cartilaginous, solid, or in age stuffed, or nearly hollow. The **veil** is floccose and quite thick when the plant is young. It is scaly on the under side, clinging to the margin of the pileus in triangular remnants, appearing like a crown. The color of the veil and of its remnants is the same as the color of the scales of the cap.

The spores in mass are light brown, and when fresh with a slight purple tinge. (The color of the spores on white paper is near walnut brown or hair brown of Ridgeway's colors.) Under the microscope they are yellowish, oval or short oblong, often inequilateral, 6–8 × 4–5 μ.

Figure 151 is from plants (No. 4016, C. U. herbarium) collected at Blowing Rock, N. C., September, 1899, on a fallen maple log. The plants sometimes occur singly. It has been collected at Ithaca, N. Y., and was first described from plants collected at Waynesville, Ohio.

Flammula sapinea Fr., is a common plant growing on dead coniferous wood. It is dull yellow, the pileus 1–4 cm. in diameter, and with numerous small scales.

HEBELOMA Fr.

In *Hebeloma* the gills are either squarely set against the stem (adnate) or they are notched (sinuate), and the spores are clay-colored. The edge of the gills is usually whitish, the surface clay-colored. The veil is only seen in the young stage, and then is very delicate and fibrillose. The stem is fleshy and fibrous, and somewhat mealy at the apex. The genus corresponds with *Tricholoma* of the white-spored agarics. All the species are regarded as unwholesome, and some are considered poisonous. The species largely occur during the autumn. Few have been studied in America. [Pg 158]

Hebeloma crustuliniforme Bull.—This plant is usually common in some of the lawns, during the autumn, at Ithaca, N. Y. It often forms rings as it grows on the ground. It is from 5–7 cm. high, the cap 4–8 cm. in diameter, and the stem is 4–6 mm. in thickness.

Figure 152.—Hebeloma crustuliniforme, var. minor. Cap whitish or tan color, or reddish-brown at center; gills clay color (natural size). Copyright.

The **pileus** is convex and expanded, somewhat umbonate, viscid when moist, whitish or tan color, darker over the center, where it is often reddish-brown. The **gills** are adnexed and rounded near the stem, crowded, whitish, then clay color and reddish-brown, the edge whitish and irregular. The **gills** are said to exude watery drops in wet weather. The **stem** is stuffed, later hollow, somewhat enlarged at the base, white, and mealy at the apex. Figure 152 is from plants (No. 2713, C. U. herbarium) collected in lawns on the Cornell University campus. The plants in this figure seem to represent the variety *minor*.

INOCYBE Fr.

In the genus *Inocybe* there is a universal veil which is fibrillose in character, and more or less closely joined with the cuticle of the pileus, and the surface of the pileus is therefore marked with fibrils or is more or less scaly. Sometimes the margin of the pileus possesses remnants of a veil which is quite prominent in a few species. The gills are adnate, or sinuate, rarely decurrent, and in one species they are free. It is thus seen that the species vary widely, and there may be, after a careful study of the species, grounds for the separation of the species into several genera. One of the most remarkable species is *Inocybe echinata* Roth. This plant is covered with a universal veil of a sooty color and powdery in nature. The gills are reddish purple, and the stem is of the same color, the spores on white paper of a faint purplish red color. Some place in it *Psalliota*. Collected at Ithaca in August, 1900.

[Pg 159] TUBARIA W. Smith.

In the genus *Tubaria* the spores are rust-red, or rusty brown (ferruginous or fuscous-ferruginous), the stem is somewhat cartilaginous, hollow, and, what is more important, the gills are more or less decurrent, broad next to the stem, and thus more or less triangular in outline. It is related to *Naucoria* and *Galera*, but differs in the decurrent gills. The pileus is convex, or with an umbilicus.

Tubaria pellucida Bull. — This species grows by roadsides in grassy places. The plants are from 3–4 cm. high, and the cap 1–2 cm. in diameter, and the stem 2–3 mm. in thickness.

Figure 153.—Tubaria pellucida. Dull reddish brown (natural size).

The **pileus** is conic, then bell-shaped, often expanded and with a slight umbo; the color is dull, reddish brown, and it has a watery appearance. The plant is sometimes enveloped with a loose and delicate universal or outer veil, which remains on the margin of the cap in the form of silky squamules as shown in the figure. The margin of the pileus is faintly striate. The **gills** are only slightly decurrent. Figure 153 is from plants (No. 2360 C. U. herbarium) collected along a street in Ithaca.

The stem is at first solid, becoming hollow, tapering above, and the apex is mealy.

CREPIDOTUS Fr.

In *Crepidotus* the pileus is lateral, or eccentric, and thus more or less shelving, or it is resupinate, that is, lying flat or nearly so on the wood. The species are usually of small size, thin, soft and fleshy. The spores are reddish brown (ferruginous). The genus corresponds to *Pleurotus* among the white-spored agarics, or to *Claudopus* among the rosy-spored ones. Peck describes eleven species in the 39th Report, N. Y. State Mus., p. 69 et seq., 1886. [Pg 160]

Crepidotus versutus Pk.—This little *Crepidotus* has a pure white pileus which is covered with a soft, whitish down. The plants grow usually on the underside of rotten wood or bark, and then the upper side of the cap lies against the wood, and is said to be resupinate. Sometimes where they grow toward the side of the log the cap has a tendency to be shelving. In the resupinate forms the cap is attached usually near one side, and then is produced more at the opposite side, so that it is more or less lateral or eccentric. As the plant becomes mature the edge is free from the wood for some distance, only being attached over a small area. The cap is somewhat reniform, thin, and from 6–12 cm. in diameter. The **gills** radiate from the point where the cap is attached to the substratum, are not crowded, rounded behind, that is, at the lateral part of the cap where they converge. They are whitish, then ferruginous from the spores. The **spores** are sub-elliptical, sometimes inequilateral, and measure from 8–12 × 4–6 μ.

Figure 154.—Crepidotus versutus. Cap white, downy; gills whitish, then rusty (twice natural size) Copyright.

Crepidotus herbarum Pk., is a closely related species, separated on account of the smaller spores. Both species grow either on herbs or decaying wood. As suggested by Peck they are both closely related to *C. chimonophilus* Berk., which has "oblong elliptical" spores. The shape of the spores does not seem to differ from the specimens which I have taken to be *C. versutus*. [Pg 161]

Crepidotus applanatus Fr., is a larger species, shelving and often imbricated. **Crepidotus fulvotomentosus** Pk., is a pretty species with a tomentose cap and tawny scales, usually occurring singly. It is closely related to *C. calolepis* Fr.

Figure 154 is from plants of *Crepidotus versutus* Pk., (No. 2732 C. U. herbarium) collected on rotting wood at Freeville, N. Y., eight miles from Ithaca. The plants are represented twice natural size.

CORTINARIUS Fr.

The genus *Cortinarius* is chiefly distinguished from the other genera of the ochre-spored agarics by the presence of a spider-web-like (arachnoid) veil which is separate from the cuticle of the pileus, that is, superficial. The gills are powdered by the spores, that is, the spores fall away with difficulty and thus give the gills a pulverulent appearance. The plants are fleshy and decay easily. It is necessary to have plants in the young as well as the old state to properly get at the characters, and the character of the veil is only seen in young or half developed specimens. The species are to be distinguished from other ochre-spored agarics with a cobwebby veil by the fact that the veil in *Cortinarius* is superficial and the gills powdery. The number of species is very large, and they are difficult to determine. They mostly occur in northern countries and in the autumn or late summer; some species, however, occur during early summer. Peck, 23d Report, N. Y. State Mus., p. 105–112, describes 21 species.

Cortinarius (Inoloma) violaceus (L.) Fr. **Edible.**—This species is known by the violet or dark violet color which pervades all parts of the plant. The plants are 8–10 cm. high, the pileus 7–15 cm. broad, and the stem is bulbous, 6–8 mm. in thickness. The veil is single. It occurs in woods and open places during late summer and in the autumn. The flesh of the plant is also violet, and this color is imparted to the liquid when the plant is cooked. The flavor is said to be something like that of *Agaricus campestris*.

Cortinarius (Myxacium) collinitus (Pers.) Fr. **Edible.**—This is known as the smeared cortinarius because of the abundant glutinous substance with which the plant is smeared during moist or wet weather. It grows in woods. The plants are 7–10 cm. high, the cap 5–8 cm. in diameter, and the stem is 8–12 mm. in thickness. It is usually known by the smooth, even, tawny cap, the great abundance of slimy substance covering the entire plant when moist, and when dry the cracking of the gluten on the stem into annular patches.

The **pileus** is convex to expanded, smooth, even, glutinous when [Pg 162] wet, shining when dry, tawny. The **gills** are adnate with a peculiar bluish gray tinge when young, and clay color to cinnamon when old. The **spores** are nearly elliptical, and 12–15 × 6–7 μ. The **stem** is cylindrical, even, and with patches of the cracked gluten when dry.

Cortinarius (Dermocybe) cinnamomeus (L.) Fr. **Edible.** — The cinnamon cortinarius is so called because of the cinnamon color of the entire plant, especially of the cap and stem. It grows in the woods during summer and autumn. It is a very pretty plant, and varies from 5–8 cm. high, the cap from 2–10 cm. broad, and the stem 4–6 mm. in thickness.

The **pileus** is conic, or convex, and nearly expanded, sometimes nearly plane, and again with a prominent blunt or conic umbo. Sometimes the pileus is abruptly bent downward near the margin as shown in the plants in Fig. 155, giving the appearance of a "hip-roof." The surface is smooth, silky, with innate fibrils. Sometimes there are cinnabar stains on parts of the pileus, and often there are concentric rows of scales near the margin. The flesh is light yellowish and with stains of cinnabar. The **gills** are adnate, slightly sinuate, and decurrent by a tooth, easily separating from the stem, rather crowded, slightly ventricose. The color of the gills varies greatly; sometimes they are the same color as the pileus, sometimes reddish brown, sometimes blood red color, etc. This latter form is a very pretty plant, and is var. *semi-sanguineus* Fr.

Figure 155. — Cortinarius cinnamomeus var. semi-sanguineus. Cap and stem cinnamon, gills blood red color (natural size). Copyright.

Figure 155 is from plants (No. 2883 C. U. herbarium) collected [Pg 163] at Ithaca. The species is widely distributed in this country as well as in Europe.

Plate 50, Figure 156.—Cortinarius ochroleucus. Entire plant pale ochre color, gills later ochre yellow (natural size). Copyright.

Cortinarius (Dermocybe) ochroleucus (Schaeff.) Fr.—This is a very beautiful plant because of the soft, silky appearance of the surface of pileus and stem, and the delicate yellowish white color. It occurs in woods, on the ground among decaying leaves. The plants are 4–12 cm. high, the cap 4–7 cm. broad, and the stem above is 6–10 mm. in thickness, and below from 2–3 cm. in thickness.

Plate 51, Figure 157.—Cortinarius ochroleucus. Colors same as in Figure 156, this represents older plants.

The **pileus** is convex to nearly expanded, and sometimes a little depressed, usually, however, remaining convex at the top. It is dry, on the center finely tomentose to minutely squamulose, sometimes the scales splitting up into concentric rows around the cap. The cap is fleshy at the center, and thin at the margin, the color is from cream buff to buff, darker on the center. The **gills** are sinuate or adnate, slightly broader in the middle (ventricose) in age, pale at first, then becoming ochre yellow, and darker when the plant dries. The **spores** are tawny in mass, oval, elliptical, minutely tuberculate when mature, 6–9 × 4–6 μ. The **stem** is clavate, pale cream buff in color, solid, becoming irregularly fistulose in age, bulbous or somewhat ventricose below, the bulb often large and abrupt, 1.5–3 cm. in diameter. The **veil** is prominent and attached to the upper part of the stem, the abundant threads attached over an area 1 cm. in extent and forming a beautiful cortina of the same color as the pileus and stem, but becoming tawny when the spores fall on it. The stem varies considerably in length and shape, being rarely ventricose, and then only at the base; the bulbous forms predominate and the bulb is often very large.

Figures 156, 157 are from plants (No. 3674 C. U. herbarium) collected at Blowing Rock, N. C., during September, 1899.

BOLBITIUS Fries.

The genus *Bolbitius* contains a few species with yellowish or yellowish brown spores. The plants are very fragile, more or less mucilaginous when moist, usually with yellowish colors, and, what is the most characteristic feature beside the yellowish color of the spores, the gills are very soft, and at maturity tend to dissolve into a mucilaginous consistency, though they do not deliquesce, or only rarely dissolve so far as to form drops. The surface of the gills at maturity becomes covered with the spores so that they appear powdery, as in the genus *Cortinarius*, which they also resemble in the color of the spores. In the mucilaginous condition of the gills the genus approaches *Coprinus*. It is believed to occupy an intermediate [Pg 164] position between *Coprinus* and *Cortinarius*. The species usually grow on dung or in manured ground, and in this respect resemble many of the species of *Coprinus*. Some of the species are, however, not always confined to such a substratum, but grow on decaying leaves, etc.

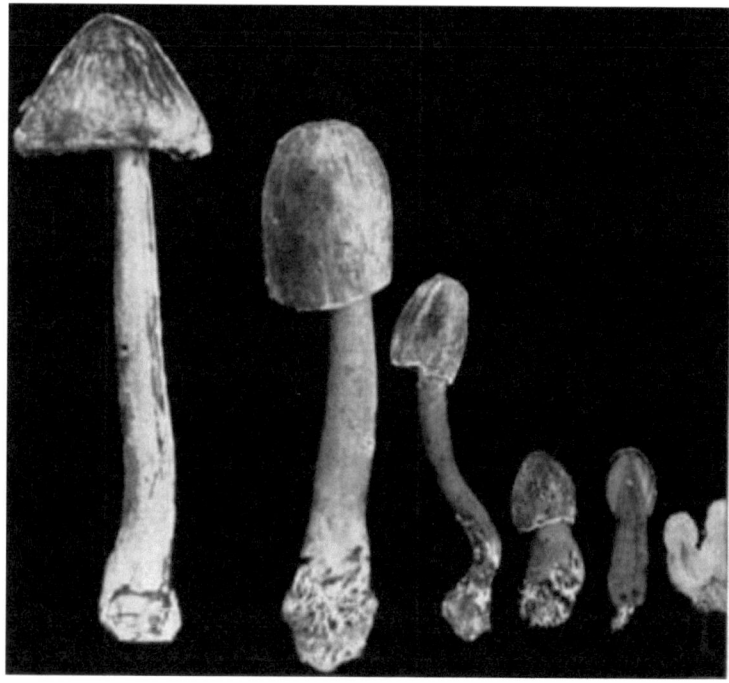

Figure 158.—Bolbitius variicolor. Cap viscid, various shades of yellow, or smoky olive; gills yellowish, then rusty (natural size).

Bolbitius variicolor Atkinson.—This plant was found abundantly during May and June, 1898, in a freshly manured grass plat between the side-walk and the pavement along Buffalo street, Ithaca, N. Y. The season was rainy, and the plants appeared each day during quite a long period, sometimes large numbers of them covering a small area, but they were not clustered nor cespitose. They vary in height from 4–10 cm., the pileus from 2–4 cm. broad, and the stem is 3–8 mm. in thickness. The colors vary from smoky to fuliginous, olive and yellow, and the spores are ferruginous.

The **pileus** is from ovate to conic when young, the margin not at all incurved, but lies straight against the stem, somewhat unequal. [Pg 165] In expanding the cap becomes convex, then expanded, and finally many of the plants with the margin elevated and with a broad umbo, and finely striate for one-half to two-thirds the way from the margin to the center. When young the pileus has a very viscid cuticle, which easily peels from the surface, showing the yellow flesh. The cuticle is smoky olive to fuliginous, darker when young, becoming paler as the pileus expands, but always darker on the umbo. Sometimes the fibres on the surface of the cap are drawn into strands which anastomose into coarse reticulations, giving the appearance of elevated veins which have a general radiate direction from the center of the cap. As the pileus expands the yellow color of the flesh shows through the cuticle more and more, especially when young, but becoming light olive to fuliginous in age. In dry weather the surface of the pileus sometimes cracks into patches as the pileus expands. The **gills** are rounded next the stem, adnate to adnexed, becoming free, first yellow, then ferruginous. The basidia are abruptly club-shaped, rather distant and separated regularly by rounded cells, four spored. The **spores** are ferruginous, elliptical, 10–15 × 6–8 μ, smooth. The **stem** is cylindrical to terete, tapering above, sulphur and ochre yellow, becoming paler and even with a light brown tinge in age. The stem is hollow, and covered with numerous small yellow floccose scales which point upward and are formed by the tearing away of the edges of the gills, which are loosely united with the surface of the stem in the young stage. The edges of the gills are thus sometimes finely fimbriate.

At maturity the gills become more or less mucilaginous, depending on the weather. Plants placed in a moist chamber change to a mucilaginous mass. When the plants dry the pileus is from a drab to hair brown or sepia color (Ridgeway's colors). Figure 158 is from plants (No. 2355 C. U. herbarium).

PAXILLUS Fr.

In the genus *Paxillus* the gills are usually easily separated from the pileus, though there are some species accredited to the genus that do not seem to possess this character in a marked degree. The spores are ochre or ochre brown. Often the gills are forked near the stem or anastomose, or they are connected by veins which them-selves anastomose in a reticulate fashion so that the meshes resem-ble the pores of certain species of the family *Polyporaceæ*. The pileus may be viscid or dry in certain species, but the plant lacks a viscid universal veil. The genus is closely related to *Gomphidius*, where the gills are often forked and easily separate from the pileus, but [Pg 166] *Gomphidius* possesses a viscid or glutinous universal veil. Peck in the Bull. N. Y. State Mus. Nat. Hist. 2: 29–33, describes five spe-cies.

Paxillus involutus (Batsch.) Fr. **Edible.**—This plant is quite common in some places and is widely distributed. It occurs on the ground in grassy places, in the open, or in woods, and on decaying logs or stumps. The stem is central, or nearly so, when growing on the ground, or eccentric when growing on wood, especially if grow-ing from the side of a log or stump. The plants are 5–7 cm. high, the cap 3–7 cm. broad, and the stem 1–2 cm. in thickness. The plant occurs from August to October.

Figure 159.—Paxillus involutus. Cap and stem gray, olive-brown, reddish brown or tawny (natural size). Copyright.

The **pileus** is convex to expanded, and depressed in the center. In the young plant the margin is strongly inrolled, and as the pileus expands it unrolls in a very pretty manner. The young plant is covered with a grayish, downy substance, and when the inrolled margin of the cap comes in contact with the gills, as it does, it presses the gills against this down, and the unrolling margin is thus marked quite prominently, sometimes with furrows where the pressure of the gills was applied. The color of the pileus varies greatly. In the case of plants collected at Ithaca and in North Carolina mountains the young plant when fresh is often olive umber, becoming reddish or tawny when older, the margin with a lighter shade. As Dr. Peck states, "it often presents a strange admixture of gray, ochraceous, ferruginous, and brown hues." The flesh is yellowish and changes [Pg 167] to reddish or brownish where bruised. The **gills** are decurrent, when young arcuate, then ascending, and are more or less reticulated on the stem. They are grayish, then greenish yellow changing to brown where bruised. The **spores** are oval, 7–9 × 4–5 µ. The **stem** is short, even, and of the same color as the cap.

Plate 52, Figure 160. — Paxillus rhodoxanthus. Cap reddish brown, stem paler, gills yellow (natural size). Copyright.

At Ithaca, N. Y., the plant is sometimes abundant in late autumn in grassy places near or in groves. The Figure 159 is from plants

(No. 2508 C. U. herbarium) growing in such a place in the suburbs of Ithaca. At Blowing Rock, N. C., the plant is often very abundant along the roadsides on the ground during August and September.

Paxillus rhodoxanthus (Schw.) — This species was first described by de Schweinitz as *Agaricus rhodoxanthus*, p. 83 No. 640, Synopsis fungorum Carolinæ superioris, in Schriften der Naturforschenden Gesellschaft 1: 19–131, 1822. It was described under his third section of *Agaricus* under the sub-genus *Gymnopus*, in which are mainly species now distributed in *Clitocybe* and *Hygrophorus*. He remarks on the elegant appearance of the plant and the fact that it so nearly resembles *Boletus subtomentosus* as to deceive one. The resemblance to *Boletus subtomentosus* as one looks upon the pileus when the plant is growing on the ground is certainly striking, because of the red-dish yellow, ochraceous rufus or chestnut brown color of the cap together with the minute tomentum covering the surface. The suggestion is aided also by the color of the gills, which one is apt to get a glimpse of from above without being aware that the fruiting surface has gills instead of tubes. But as soon as the plant is picked and we look at the under surface, all suggestion of a *Boletus* vanishes, unless one looks carefully at the venation of the surface of the gills and the spaces between them. The plant grows on the ground in woods. At Blowing Rock, N. C., where it is not uncommon, I have always found it along the mountain roads on the banks. It is 5–10 cm. high, the cap from 3–8 cm. broad, and the stem 6–10 mm. in thickness.

The **pileus** is convex, then expanded, plane or convex, and when mature more or less top-shaped because it is so thick at the middle. In age the surface of the cap often becomes cracked into small areas, showing the yellow flesh in the cracks. The flesh is yellowish and the surface is dry. The **gills** are not very distant, they are stout, chrome yellow to lemon yellow, and strongly decurrent. A few of them are forked toward the base, and the surface and the space between them are marked by anastomosing veins forming a reticulum suggestive of the hymenium of the *Polyporaceæ*. This character is [Pg 168] not evident without the use of a hand lens. The surface of the gills as well as the edges is provided with clavate **cystidia** which are filled with a yellow pigment, giving to the gills the bright yellow color so characteristic. These cystidia extend above the basidia, and the ends are rounded so that sometimes they appear capitate. The yellow color is not confined to the cystidia, for the sub-hymenium is also colored in a similar way. The **spores** are yellowish, oblong to elliptical or spindle-shaped, and measure 8–12 × 3–5 μ. The **stem** is the same color as the pileus, but paler, and more yellow at the base. It is marked with numerous minute dots of a darker color than the ground color, formed of numerous small erect tufts of mycelium.

Figure 160 is from plants (No. 3977 C. U. herbarium) collected at Blowing Rock, N. C., during September, 1899. As stated above, the plant was first described by de Schweinitz as *Agaricus rhodoxanthus* in 1822. In 1834 (Synop. fung. Am. Bor. p. 151, 1834) he listed it under the genus *Gomphus* Fries (Syst. Mycolog. 319, 1821). Since Fries changed *Gomphus* to *Gomphidius* (Epicrisis, 319, 1836–1838) the species has usually been written *Gomphidius rhodoxanthus* Schweinitz. The species lacks one very important characteristic of the genus *Gomphidius*, namely, the slimy veil which envelops the entire plant. Its relationship seems rather to be with the genus *Paxillus*, though the gills do not readily separate from the pileus, one of the characters ascribed to this genus, and possessed by certain species of *Gomphidius* in even a better degree. (In Paxillus involutus the gills do not separate so readily as they do in certain species of *Gomphidius*.) Berkeley (Decades N. A. Fungi, 116) has described a plant from Ohio under the name *Paxillus flavidus*. It has been suggested by some (see Peck, 29th Report, p. 36; Lloyd, Mycolog. Notes, where he writes it as *Flammula rhodoxanthus*!) that *Paxillus flavidus* Berk., is identical with *Agaricus rhodoxanthus* Schw.

Paxillus rhodoxanthus seems also to be very near if not identical with *Clitocybe pelletieri* Lev. (Gillet, Hymenomycetes **1**: 170), and Schroeter (Cohn's Krypt, Flora Schlesien, **3**, 1: 516, 1889) transfers this species to *Paxillus* as *Paxillus pelletieri*. He is followed by Hennings, who under the same section of the genus, lists *P. flavidus* Berk., from N. A. The figure of *Clitocybe pelletieri* in Gillet Hymenomycetes, etc., resembles our plant very closely, and Saccardo (Syll. Fung. **5**: 192) says that it has the aspect of *Boletus subtomentosus,* a remark similar to the one made by de Schweinitz in the original description of *Agaricus rhodoxanthus*. *Flammula paradoxa* Kalch. (Fung. Hung. Tab. XVII, Fig. 1) seems to be the same plant, as [Pg

169] well as *F. tammii* Fr., with which Patouillard (Tab. Anal. N. 354) places *F. paradoxa* and *Clitocybe pelletieri*.

Paxillus atro=tomentosus (Batsch) Fr.—This plant is not very common. It is often of quite large size, 6–15 cm. high, and the cap 5–10 cm. broad, the stem very short or sometimes long, from 1–2.5 cm. in thickness. The plant is quite easily recognized by the stout and black hairy stem, and the dark brown or blackish, irregular and sometimes lateral cap, with the margin incurved. It grows on wood, logs, stumps, etc., during late summer and autumn.

Figure 161.—Paxillus atro-tomentosus, form hirsutus. Cap and stem brownish or blackish (natural size, small specimens, they are often larger). Copyright.

The **pileus** is convex, expanded, sometimes somewhat depressed, lateral, irregular, or sometimes with the stem nearly in the center, brownish or blackish, dry, sometimes with a brownish or blackish tomentum on the surface. The margin is inrolled and later incurved. The flesh is white, and the plant is tough. The **gills** are adnate, often decurrent on the stem, and easily separable from the pileus, forked at the base and sometimes reticulate, forming pores. **Spores** yellowish, oval, 4–6 × 3–4 μ. Stevenson says that the gills do not form pores like those of P. involutus, but Fig. 161 (No. 3362 C. U. herbarium) from plants collected at Ithaca, shows them well. There is, [Pg 170] as it seems, some variation in this respect. The **stem** is solid, tough and elastic, curved or straight, covered with a dense black tomentum, sometimes with violet shades. On drying the plant becomes quite hard, and the gills blackish olive.

Paxillus panuoides Fr. — This species was collected during August, 1900, on a side-walk and on a log at Ithaca. The specimens collected were sessile and the **pileus** lateral, somewhat broadened at the free end, or petaloid. The entire plant is pale or dull yellow, the surface of the pileus fibrous and somewhat uneven but not scaly. The plants are 2–12 cm. long by 1–8 cm. broad, often many crowded together in an imbricated manner. The **gills** are pale yellow, and the **spores** are of the same color when caught on white paper, and they measure 4–5 × 3–4 μ, the size given for European specimens of this species. The gills are forked, somewhat anastomosing at the base, and sinuous in outline, though not markedly corrugated as in the next form. From descriptions of the European specimens the plants are sometimes larger than these here described, and it is very variable in form and often imbricated as in the following species.

Paxillus corrugatus Atkinson. — This very interesting species was collected at Ithaca, N. Y., on decaying wood, August 4, 1899. The pileus is lateral, shelving, the stem being entirely absent in the specimens found. The **pileus** is 2–5 cm. broad, narrowed down in an irregular wedge form to the sessile base, convex, then expanded, the margin incurved (involute). The color of the cap is yellow, maize yellow to canary yellow, with a reddish brown tinge near the base. It is nearly smooth, or very slightly tomentose. The flesh is pale yellow, spongy. The **gills** are orange yellow, 2–3 mm. broad, not crowded, regularly forked several times, thin, blunt, very wavy and crenulate, easily separating from the hymenophore when fresh; the entire breadth of the gills is fluted, giving a corrugated appearance to the side. The **spores** in these specimens are faintly yellow, minute, oblong, broadly elliptical, short, sometimes nearly oval, 3 × 1.5–2 μ. The **basidia** are also very minute. The spores are olive yellow on white paper. The plant has a characteristic and disagreeable odor. This odor persists in the dried plant for several months.

Figure 162 is from the plants (No. 3332 C. U. herbarium) collected as noted above on decaying hemlock logs in woods. A side and under view is shown in the figure, and the larger figure is the under-view, from a photograph made a little more than twice natural size, in order to show clearly the character of the gills. The two smaller plants are natural size. When dry the plant is quite hard.

Plate 53, Figure 162.—Paxillus corrugatus. Cap maize yellow to orange yellow, reddish brown near the base; gills orange yellow. Two lower plants natural size; upper one 2-1/2 times natural size. Copyright.

Plate 54, Figure 163. — Paxillus panuoides, pale yellow; natural size. Copyright.

- PLATE 55.
- Fig. 1. — Boletus felleus.
- Fig. 2. — B. edulis.
- Copyright 1900.

CHAPTER IX. [Pg 171]

THE TUBE-BEARING FUNGI. POLYPORACEAE.

The plants belonging to this family are characterized especially by a honey-combed fruiting surface, that is, the under surface of the plants possesses numerous tubes or pores which stand close together side by side, and except in a very few forms these tubes are joined by their sides to each other. In *Fistulina* the tubes are free from each other though standing closely side by side. In *Merulius* distinct tubes are not present, but the surface is more or less irregularly pitted, the pits being separated from each other by folds which anastomose, forming a network. These pits correspond to shallow tubes.

The plants vary greatly in consistency, some are very fleshy and soft and putrify readily. Others are soft when young and become firmer as they age, and some are quite hard and woody. Many of the latter are perennial and live for several or many years, adding a new layer in growth each year. The larger number of the species grow on wood, but some grow on the ground; especially in the genus *Boletus*, which has many species, the majority grow on the ground. Some of the plants have a cap and stem, in others the stem is absent and the cap attached to the tree or log, etc., forms a shelf, or the plant may be thin and spread over the surface of the wood in a thin patch.

In the genus *Dædalea* the tubes become more or less elongated horizontally and thus approach the form of the gills, while in some species the tubes are more or less toothed or split and approach the spine-bearing fungi at least in appearance of the fruit-bearing surface. Only a few of the genera and species will be described.

The following key is not complete, but may aid in separating some of the larger plants:

	Tubes or pores free from each other, though standing closely side by side,	*Fistulina.*
	Tubes or pores not free, joined side by side,	1.
1 —	Plants soft and fleshy, soon decaying,	2.

	Plants soft when young, becoming firm, some woody or corky, stipitate, shelving, or spread over the wood,	*Polyporus.*
[Pg 172]	Tubes or pores shallow, formed by a network of folds or wrinkles, plants thin, sometimes spread over the wood, and somewhat gelatinous,	*Merulius.*
2 —	Mass (stratum) of tubes easily separating from the cap when peeled off, cap not with coarse scales, tubes in some species in radiating lines,	*Boletus.*
	Stratum of tubes separating, but not easily, cap with coarse, prominent scales,	*Strobilomyces.*
	Stratum of tubes separating, but not easily, tubes arranged in distinct radiating lines. In one species (*B. porosus*) the tubes do not separate from the cap,	*Boletinus.*

This last genus is apt to be confused with certain species of Boletus which have a distinct radiate arrangement of the tubes. It is questionable whether it is clearly distinguished from the genus Boletus.

BOLETUS Dill.

Of the few genera in the *Polyporaceæ* which are fleshy and putrescent, *Boletus* contains by far the largest number of species. The entire plant is soft and fleshy, and decays soon after maturity. The stratum of tubes on the under side of the cap is easily peeled off and separates as shown in the portion of a cap near the right hand side of Fig. 169. In the genus *Polyporus* the stratum of tubes cannot thus be separated. In the genera *Strobilomyces* and *Boletinus*, two other fleshy genera of this family, the separation is said to be more difficult than in *Boletus*, but it has many times seemed to me a "distinction without a difference."

The larger number of the species of *Boletus* grow on the ground. Some change color when bruised or cut, so that it is important to note this character when the plant is fresh, and the taste should be noted as well.

Boletus edulis Bull. **Edible.** [*Ag. bulbosus* Schaeff. Tab. 134, 1763. *Boletus bulbosus* (Schaeff.) Schroeter. Cohn's Krypt, Flora. Schlesien, p. 499, 1889].—This plant, which, as its name implies, is edible, grows in open woods or their borders, in groves and in open places, on the ground. It occurs in warm, wet weather, from July to September. It is one of the largest of the Boleti, and varies from 5–12 cm. high, the cap from 8–25 cm. broad, and the stem 2–4 cm. in thickness.

Plate 56, Figure 164.—Boletus edulis. Cap light brown, tubes greenish yellow or yellowish; stem in this specimen entirely reticulate (natural size, often larger). Copyright.

The **pileus** is convex to expanded, smooth, firm, quite hard when young and becoming soft in age. The color varies greatly, from [Pg 173] buff to dull reddish, to reddish-brown, tawny-brown, often yellowish over a portion of the cap, usually paler on the margin. The flesh is white or tinged with yellow, sometimes reddish under the cuticle. The **tubes** are white when young and the mouths are closed (stuffed), the lower surface of the tubes is convex from the margin of the cap to the stem, and depressed around the stem, sometimes separating from the stem. While the tubes are white when young, they become greenish or greenish-yellow, or entirely yellow when mature. The **spores** when caught on paper are greenish-yellow, or yellow. They are oblong to fusiform, 12–15 µ long. The **stem** is stout, even, or much enlarged at the base so that it is clavate. The surface usually shows prominent reticulations on mature plants near the tubes, sometimes over the entire stem. This is well shown in Fig. 164 from plants (No. 2886, C. U. herbarium) collected at Ithaca, N. Y.

Plate 57, Figure 165.—Cap light brown, tubes greenish yellow or yellowish; stem in these specimens not reticulate (2/3 natural size). Copyright.

Figure 165 represents plants (No. 4134, C. U. herbarium) collected at Blowing Rock, N. C., in September, 1899. The plant is widely distributed and has long been prized as an esculent in Europe and America. When raw the plant has an agreeable nutty taste, sometimes sweet. The caps are sometimes sliced and dried for future use. It is usually recommended to discard the stems and remove the tubes since the latter are apt to form a slimy mass on cooking.

Boletus felleus Bull. **Bitter.** — This is known as the bitter boletus, because of a bitter taste of the flesh. It usually grows on or near much decayed logs or stumps of hemlock spruce. It is said to be easily recognized by its bitter taste. I have found specimens of a plant which seems to have all the characters of this one growing at the base of hemlock spruce trees, except that the taste was not bitter. At Ithaca, however, the plant occurs and the taste is bitter. It is one of the large species of the genus, being from 8–12 cm. high, the cap 7–20 cm. broad, and the stem 1–2.5 cm. in thickness.

The **pileus** is convex becoming nearly plane, firm, and in age soft, smooth, the color varying from pale yellow to various shades of brown to chestnut. The flesh is white, and where wounded often changes to a pink color, but not always. The **tubes** are adnate, long, the under surface convex and with a depression around the stem. The tubes are at first white, but become flesh color or tinged with flesh color, and the mouths are angular. The **stem** is stout, tapering upward, sometimes enlarged at the base, usually reticulated at the upper end, and sometimes with the reticulations over the entire surface (Fig. 166). The color is paler than that of the cap. The **spores** are oblong to spindle-shaped, flesh color in mass, and single ones measure 12–18 × 4–5 μ. [Pg 174]

The general appearance of the plant is somewhat like that of the *Boletus edulis*, and beginners should be cautioned not to confuse the two species. It is known by its bitter taste and the flesh-colored tubes, while the taste of the *B. edulis* is sweet, and the tubes are greenish-yellow, or yellowish or light ochre.

Plate 55 represents three specimens in color.

Boletus scaber Fr. **Edible.** — This species is named the rough-stemmed boletus, in allusion to the rough appearance given to the stem from numerous dark brown or reddish dots or scales. This is a characteristic feature, and aids one greatly in determining the species, since the color of the cap varies much. The cap is sometimes whitish, orange red, brown, or smoky in color. The plant is 6–15 cm. high, the cap 3–7 cm. broad, and the stem 8–12 mm. in thickness.

The **pileus** is rounded, becoming convex, smooth, or nearly so, sometimes scaly, and the flesh is soft and white, sometimes turning slightly to a reddish or dark color where bruised. The **tubes** are small, long, the surface formed by their free ends is convex in outline, and the tubes are depressed around the stem. They are first white, becoming darker, and somewhat brownish. The **stem** is solid, tapering somewhat upward, and roughened as described above.

The plant is one of the common species of the genus *Boletus*. It occurs in the woods on the ground or in groves or borders of woods in grassy places. Writers differ as to the excellence of this species for food; some consider it excellent, while others regard it as less agreeable than some other species. It is, at any rate, safe, and Peck considers it "first-class."

Boletus retipes B. & C. — This species was first collected in North Carolina by Curtis, and described by Berkeley. It has since been reported from Ohio, Wisconsin, and New England (Peck, Boleti of the U. S.). Peck reported it from New York in the 23d Report, N. Y. State Mus., p. 132. Later he recognized the New York plant as a new species which he called *B. ornatipes* (29th Report, N. Y. State Mus., p. 67). I collected the species in the mountains of North Carolina, at Blowing Rock, in August, 1888. During the latter part of August and in September, 1899, I had an opportunity of seeing quite a large number of specimens in the same locality, for it is not uncommon there, and two specimens were photographed and are represented here in Fig. 167. The original description published in Grevillea **1**: 36, should be modified, especially in regard to the size of the plant, its habit, and the pulverulent condition of the pileus. The plants are 6–15 cm. high, the cap 5–10 cm. broad, and the stem 0.5–1.5 cm. in thickness.

[Pg 175]

Plate 58, Figure 166. — Boletus felleus. Cap light brown, tubes flesh color, stem in this specimen entirely reticulate (natural size, often larger). Copyright.

Figure 167.—Boletus retipes. Cap yellowish brown, to olive-brown or nearly black, stem yellow, beautifully reticulate, tubes yellow (natural size). Copyright.

The **pileus** is convex, thick, soft and somewhat spongy, especially in large plants. The cap is dry and sometimes, especially when young, it is powdery; at other times, and in a majority of cases according to my observations, it is not powdery. It is smooth or minutely tomentose, sometimes the surface cracked into small patches, but usually even. The color varies greatly between yellowish brown to olive brown, fuliginous or nearly black. The **tubes** are yellow, adnate, the tube surface plane or convex. The spores are yellowish or ochraceous, varying somewhat in tint in different specimens. The **stem** is yellow, yellow also within, and beautifully reticulate, [Pg

176] usually to the base, but sometimes only toward the apex. It is usually more strongly reticulate over the upper half. The stem is erect or ascending.

The plant grows in woods, in leaf mold or in grassy places. It is usually single, that is, so far as my observations have gone at Blowing Rock. Berkeley and Curtis report it as cespitose. I have never seen it cespitose, never more than two specimens growing near each other.

Boletus ornatipes Pk., does not seem to be essentially different from *B. retipes*. Peck says (Boleti U. S., p. 126) that "the tufted mode of growth, the pulverulent pileus and paler spores separate this species" (*retipes*) "from the preceding one" (*ornatipes*). Inasmuch as I have never found *B. retipes* tufted, and the fact that the pileus is not always pulverulent (the majority of specimens I collected were not), and since the tint of the spores varies as it does in some other species, the evidence is strong that the two names represent two different habits of the same species. The tufted habit of the plants collected by Curtis, or at least described by Berkeley, would seem to be a rather unusual condition for this species, and this would account for the smaller size given to the plants in the original description, where the pileus does not exceed 5 cm. in diameter, and the stem is only 5 cm. long, and 6-12 mm. in thickness. Plants which normally occur singly do on some occasions occur tufted, and then the habit as well as the size of the plant is often changed.

A good illustration of this I found in the case of *Boletus edulis* during my stay in the North Carolina mountains. The plant usually occurs singly and more or less scattered. I found one case where there were 6-8 plants in a tuft, the caps were smaller and the stems in this case considerably longer than in normal specimens. A plant which agrees with the North Carolina specimens I have collected at Ithaca, and so I judge that *B. retipes* occurs in New York.

Boletus chromapes Frost. — This is a pretty boletus, and has been reported from New England and from New York State. During the summer of 1899 it was quite common in the Blue Ridge mountains, North Carolina. The plant grows on the ground in woods. It is 6–10 cm. high, the cap is 5–10 cm. in diameter, and the stem is 8–12 mm.

in thickness. It is known by the yellowish stem covered with red-dish glandular dots.

Plate 59, Figure 168.—Boletus chromapes. Cap pale red, rose or pink, tubes flesh color, then brown, stem yellowish either above or below, the surface with reddish or pinkish dots (natural size). Copyright.

The **pileus** is convex to nearly expanded, pale red, rose pink to vinaceous pink in color, and sometimes slightly tomentose. The flesh is white, and does not change when cut or bruised. The **tube** surface is convex, and the tubes are attached slightly to the stem, [Pg 177] or free. They are white, then flesh color, and in age become brown. The **stem** is even, or it tapers slightly upward, straight or ascending, whitish or yellow above, or below, sometimes yellowish the entire length. The flesh is also yellowish, especially at the base. The entire surface is marked with reddish or pinkish dots.

Plate 60, Figure 169.—Boletus vermiculosus. Cap brown to gray or buff; tubes yellowish with reddish brown mouths; flesh quickly changes to blue where wounded (natural size, sometimes larger). Copyright.

Figure 168 is from plants (No. 4085 C. U. herbarium) collected at Blowing Rock, N. C., during September, 1899.

Boletus vermiculosus Pk.—This species was named *B. vermiculosus* because it is sometimes very "wormy." This is not always the case, however. It grows in woods on the ground, in the Eastern United States. It is from 6–12 cm. high, the cap from 7–12 cm. broad, and the stem 1–2 cm. in thickness.

The **pileus** is thick, convex, firm, smooth, and varies in color from brown to yellowish brown, or drab gray to buff, and is minutely tomentose. The flesh quickly changes to blue where wounded, and the bruised portion, sometimes, changing to yellowish. The **tubes** are yellowish, with reddish-brown mouths, the tube surface being rounded, free or nearly so, and the tubes changing to blue where wounded. The **stem** is paler than the pileus, often dotted with short, small, dark tufts below, and above near the tubes abruptly paler, and sometimes the two colors separated by a brownish line. The stem is not reticulated. Figure 169 is from a photograph of plants (No. 4132 C. U. herbarium) collected at Blowing Rock, N. C., during September, 1899.

Boletus obsonium (Paul.) Fr.—This species was not uncommon in the woods at Blowing Rock, N. C., during the latter part of August and during September, 1899. It grows on the ground, the plants usually appearing singly. It is from 10–15 cm. high, the cap 8–13 cm. broad, and the stem 1–2 cm. in thickness, considerably broader at the base than at the apex.

The **pileus** is convex to expanded, vinaceous cinnamon, to pinkish vinaceous or hazel in color. It is soft, slightly tomentose, and when old the surface frequently cracks into fine patches showing the pink flesh beneath. The thin margin extends slightly beyond the tubes, so that it is sterile. The flesh does not change color on exposure to the air. The **tubes** are plane, adnate, very slightly depressed around the stem or nearly free, yellowish white when young, becoming dark olive green in age from the color of the spores. The tube mouths are small and rotund. The **spores** caught on white paper are dark olive green. They are elliptical usually, with rounded ends, 12–15 × 4–5 μ. The **stem** is white when young, with a tinge of yellow ochre, and pale flesh color below. It is marked with somewhat parallel [Pg 178] elevated lines, or rugæ below, where it is enlarged and nearly bulbous. In age it becomes flesh color the entire length and is more plainly striate rugose with a yellowish tinge at the base. The stem tapers gradually and strongly from the base to the apex, so that it often appears long conic.

The plant is often badly eaten by snails, so that it is sometimes difficult to obtain perfect specimens. Figure 170 is from a photograph of plants (No. 4092 C. U. herbarium) from Blowing Rock, N. C.

Boletus americanus Pk.—This species occurs in woods and open places, growing on the ground in wet weather. It occurs singly or clustered, sometimes two or three joined by their bases, but usually more scattered. It is usually found under or near pine trees. The plant is 3–6 cm. high, the cap 2–7 cm. broad, and the stem is 4–8 mm. in thickness. It is very slimy in wet weather, the cap is yellow, streaked or spotted with faint red, and the stem is covered with numerous brown or reddish brown dots.

The **pileus** is rounded, then convex, becoming nearly expanded and sometimes with an umbo. It is soft, very slimy or viscid when moist, yellow. When young the surface gluten is often mixed with loose threads, more abundant on the margin, and continuous with the veil, which can only be seen in the very young stage. As the pileus expands the margin is sometimes scaly from remnants of the veil and of loose hairs on the surface. The cap loses its bright color as it ages, and is then sometimes streaked or spotted with red. The **tube** surface is nearly plane, and the tubes join squarely against the stem. The tubes are rather large, angular, yellowish, becoming dull ochraceous. The **stem** is nearly equal, yellow, and covered with numerous brownish or reddish brown glandular dots. No ring is present.

This species grows in the same situations as the *B. granulatus*, sometimes both species are common over the same area. Figure 171 is from plants (No. 3991 C. U. herbarium) collected at Blowing Rock, N. C., September, 1899. The species is closely related to *B. flavidus* Fr., and according to some it is identical with it.

Boletus granulatus L. **Edible.**—This species is one of the very common and widely distributed ones. It grows in woods and open places on the ground. Like *B. americanus*, it is usually found under or near pines. It occurs during the summer and autumn, sometimes appearing very late in the season. The plants are 3–6 cm. high, the cap is 4–10 cm. broad, and the stem is 8–12 mm. in thickness. The plants usually are clustered, though not often very crowded.

[Pg 179]

Plate 61, Figure 170.—Boletus obsonium. Cap cinnamon to pink or hazel in color, slightly tomentose; stem white, then pale flesh color (natural size). Copyright.

The **pileus** is convex to nearly expanded, flat. When moist it is very viscid and reddish brown, paler and yellowish when it is dry, but very variable in color, pink, red, yellow, tawny, and brown shades. The flesh is pale yellow. The **tubes** are joined squarely to the stem, short, yellowish, and the edges of the tubes, that is, at the open end (often called the mouth), are dotted or granulated. The **stem** is dotted in the same way above. The **spores** in mass are pale yellow; singly they are spindle-shaped.

Figure 171.—Boletus americanus. Cap slimy, yellow, sometimes with reddish spots, tubes yellowish (natural size). Copyright.

The species is edible, though some say it should be regarded with suspicion. Peck has tried it, and I have eaten it, but the viscid character of the plant did not make it a relish for me. There are several species closely related to the granulated Boletus. *B. brevipes* Pk., is one chiefly distinguished by the short stem, which entirely lacks the glandular dots. It grows in sandy soil, in pine groves and in woods.

Boletus punctipes Pk.—This species has been reported from New York State by Peck. During [Pg 180] September, 1899, I found it quite common in the Blue Ridge mountains of North Carolina, at an elevation of between 4000 and 5000 feet. It grows on the ground in mixed woods. The plants are 5–8 cm. high, the caps 5–7 cm. broad, and the stem 6–10 mm. in thickness.

Figure 172.—Boletus punctipes. Cap viscid when moist, reddish brown, pink, yellow, tawny, etc., tubes yellowish, stem dark punctate (natural size). Copyright.

The **pileus** is convex, sometimes becoming nearly plane, and it is quite thick in the center, more so than the granulated boletus, while the margin is thin, and when young with a minute gray powder. The margin often becomes upturned when old; the cap is viscid when moist, dull yellow. The **tubes** are short, their lower surface plane, and they are set squarely against the stem. They are small, the mouths rounded, brownish, then dull ochraceous, and dotted with glandules. The **stem** is rather long, proportionately more so than in the granulated boletus. It distinctly tapers upwards, is "rhubarb yellow," and dotted with glandules. This character of the stem suggested the name of the species. The **spores** are 8–10 × 4–5 μ. Figure 172 is from plants (No. 4067 C. U. herbarium) collected at Blowing Rock, N. C. It is closely related to *B. granulatus* and by some is considered the same. [Pg 181]

Boletus luteus Linn. (*B. subluteus* Pk.) This species is widely distributed in Europe and America, and grows in sandy soil, in pine or mixed woods or groves. The plants are 5–8 cm. high, the cap 3–12 cm. in diameter, and the stem 6–10 mm. in thickness. The general color is dull brown or yellowish brown, and the plants are slimy in moist weather, the stem and tubes more or less dotted with dark points. These characters vary greatly under different conditions, and the fact has led to some confusion in the discrimination of species.

Figure 173.—Boletus luteus. Cap viscid when moist, dull yellowish to reddish brown, tubes yellowish, stem punctate both above and below the annulus (natural size). Copyright.

The **pileus** is convex, becoming nearly plane, viscid or glutinous when moist, dull yellowish to reddish brown, sometimes with the color irregularly distributed in streaks. The flesh is whitish or dull yellowish. The **tube** surface is plane or convex, the tubes set squarely against the stem (adnate), while the tubes are small, with small, nearly rounded, or slightly angular mouths. The color of the tubes is yellowish or ochre colored, becoming darker in age, and sometimes nearly brown or quite dark. The **stem** is pale yellowish, reddish or brownish, and more or less covered with glandular dots, which when dry give a black dotted appearance to the stem. In the case of descriptions of *B. luteus* the stem is said to be dotted only above the annulus, while the description of *B. subluteus* gives the stem as dotted both above and below the annulus. [Pg 182] The **spores** are yellowish brown or some shade of this color in mass, lighter yellowish brown under the microscope, fusiform or nearly so, and 7–10 × 2–4 µ. The **annulus** is very variable, sometimes collapsing as a narrow ring around the stem as in Fig. 173, from plants collected at Blowing Rock, N. C., September, 1899 (*B. subluteus* Pk.), and sometimes appearing as a broad, free collar, as in Fig. 174. The veil is more or less gelatinous, and in an early stage of the plant may cover the stem as a sheath. The lower part of the stem is sometimes covered at maturity with the sheathing portion of the veil, the upper part only appearing as a ring. In this way, the lower part of the stem being covered, the glandular dots are not evident, while the stem is seen to be dotted above the annulus. But in many cases the veil slips off from the lower portion of the stem at an early stage, and then in its slimy condition collapses around the upper part of the stem, leaving the stem uncovered and showing the dots both above and below the ring (*B. subluteus*).

Plate 62, Figure 174.—Boletus luteus. Cap drab to hair-brown with streaks of the latter, viscid when moist, tubes tawny olive to walnut-brown, stem black dotted both above and below the broad, free annulus (natural size). Copyright.

An examination of the figures of the European plant shows that the veil often slips off from the lower portion of the stem in *B. luteus*, especially in the figures given by Krombholtz, T. 33. In some of these figures the veil forms a broad, free collar, and the stem is then dotted both above and below, as is well shown in the figures. In other figures where the lower part of the veil remains as a sheath over the lower part of the stem, the dots are hidden. I have three specimens of the *B. luteus* of Europe from Dr. Bresadola, collected at Trento, Austria-Hungary: one of them has the veil sheathing the lower part of the stem, and the stem only shows the dots above the annulus; a second specimen has the annulus in the form of a collapsed ring near the upper end of the stem, and the stem dotted both above and below the annulus; in the third specimen the annulus is in the form of a broad, free collar, and the stem dotted both above and below. The plants shown in Fig. 174 (No. 4124, C. U. herbarium) were collected at Blowing Rock, N. C., during September, 1899. They were found in open woods under Kalmia where the sun had an opportunity to dry out the annulus before it became collapsed or agglutinated against the stem, and the broad, free collar was formed. My notes on these specimens read as follows: "The **pileus** is

293

convex, then expanded, rather thick at the center, the margin thin, sometimes sterile, incurved. In color it runs from ecru drab to hair-brown with streaks of the latter, and it is very viscid when moist. When dried the surface of the pileus is shining. The **tubes** are plane or concave, adnate, tawny-olive to walnut-brown. The tubes are [Pg 183] small, angular, somewhat as in *B. granulatus*, but smaller, and they are granulated with reddish or brownish dots. The **spores** are walnut brown, oblong to elliptical, 8–10 × 2–3 µ. The **stem** is cylindrical, even, olive yellow above, and black dotted both above and below the annulus."

Figure 175.—Boletinus pictus. Cap reddish, tinged with yellowish between the scales, stem same color, tubes yellow, often changing to reddish brown where bruised (natural size). Copyright.

Boletinus pictus Pk.—This very beautiful plant is quite common in damp pine woods. It is easily recognized by the reddish cottony layer of mycelium threads which cover the entire plant when y-oung, and form a veil which covers the gills at this time. As the plant expands the reddish outer layer is torn into scales of the same color, showing the yellowish, or pinkish, flesh beneath, and the flesh often changes to pink or reddish where wounded. The tubes are first pale yellow, but become darker in age, often changing to pinkish, with a brown tinge where bruised. The stem is solid, and is thus different from a closely related species, *B. cavipes* Kalchb. The stem is covered with a coat like that on the pileus and is similarly colored, though often paler. The spores are [Pg 184] ochraceous, 15–18 × 6–8 µ. The plants are 5–8 cm. high, the caps 5–8 cm. broad, and the stems 6–12 mm. in thickness.

Figure 175 is from plants collected in the Blue Ridge mountains, Blowing Rock, N. C., September, 1899.

Boletinus porosus (Berk.) Pk.—This very interesting species is widely distributed in the Eastern United States. It resembles a *Polyporus*, though it is very soft like a *Boletus*, but quite tenacious. The plants are dull reddish-brown, viscid when moist, and shining. The cap is more or less irregular and the stem eccentric, the cap being sometimes more or less lobed. The plants are 4–6 cm. high, the cap 5–12 cm. broad, and the short stem 8–12 mm. in thickness. It occurs in damp ground in woods.

The **pileus** is fleshy, thick at the middle, and thin at the margin. The **tubes** are arranged in prominently radiating rows, the partitions often running radiately in the form of lamellæ, certain ones of them being more prominent than others as shown in Fig. 176. These branch and are connected by cross partitions of less prominence. This character of the hymenium led Berkeley to place the plant in the genus *Paxillus*, with which it does not seem to be so closely related as with the genus *Boletus*. The stratum of tubes, though very soft, is very tenacious, and does not separate from the flesh of the pileus, thus resembling certain species of *Polyporus*. Figure 176 is from plants collected at Ithaca.

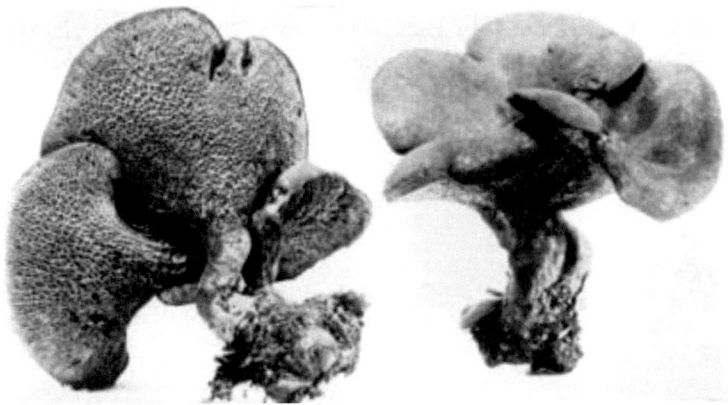

Plate 63, Figure 176.—Boletinus porosus. Viscid when moist, dull reddish brown (natural size). Copyright.

Strobilomyces strobilaceus Berk. **Edible.** — This plant has a peculiar name, both the genus and the species referring to the cone-like appearance of the cap with its coarse, crowded, dark brown scales, bearing a fancied resemblance to a pine cone. It is very easily distinguished from other species of *Boletus* because of this character of the cap. The plant has a very wide distribution though it is not usually very common. The plant is 8–14 cm. high, the cap 5–10 cm. broad, and the stem 1–2 cm. in thickness.

The **pileus** is hemispherical to convex, shaggy from numerous large blackish, coarse, hairy, projecting scales. The margin of the cap is fringed with scales and fragments of the veil which covers the tubes in the young plants. The flesh is whitish, but soon changes to reddish color, and later to black where wounded or cut. The **tubes** are adnate, whitish, becoming brown and blackish in the older plants. The mouths of the tubes are large and angular, and change color where bruised, as does the flesh of the cap. The stem is even, or sometimes tapers upward, often grooved near the apex, very tomentose or scaly with soft scales of the same color as the cap. The **spores** are in mass dark brown, nearly globose, roughened, and [Pg 185] 10–12 μ long. Figs. 177–179 are from plants collected at Ithaca, N. Y. Another European plant, *S. floccopus* Vahl, is said by Peck to occur in the United States, but is much more rare. The only difference in the two noted by Peck in the case of the American plants is that the tubes are depressed around the stem in *S. floccopus*.

Plate 64, Figure 177.—Strobilomyces strobilaceus. Scales of cap dark brown or black, flesh white but soon changing to reddish and later to black where wounded, stem same color but lighter (natural size). Copyright.

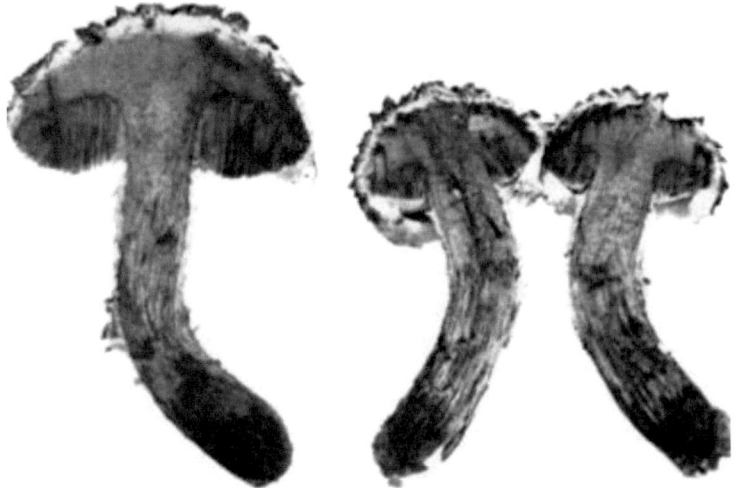

Figure 178.—Strobilomyces strobilaceus. Sections of plants. Copyright.

Figure 179.—Strobilomyces strobilaceus. Under view. Copyright.

FISTULINA Bull. [Pg 186]

In the genus *Fistulina* the tubes, or pores, are crowded together, but stand separately, that is, they are not connected together, or grown together into a stratum as in *Boletus* and other genera of the family *Polyporaceæ*. When the plant is young the tubes are very short, but they elongate with age.

Fistulina hepatica Fr. **Edible.**—This is one of the largest of the species in the genus and is the most widely distributed and common one. It is of a dark red color, very soft and juicy. It has usually a short stem which expands out into the broad and thick cap. When young the upper side of the cap is marked by minute elevations of a different color, which suggest the papillæ on the tongue; in age the tubes on the under surface have also some such suggestive appearance. The form, as it stands outward in a shelving fashion from stumps or trees, together with the color and surface characters, has suggested several common names, as beef tongue, beef-steak fungus, oak or chestnut tongue. The plant is 10–20 cm. long, and 8–15 cm. broad, the stem very short and thick, sometimes almost wanting, and again quite long. I have seen some specimens growing from a hollow log in which the stems were 12–15 cm. long.

The **pileus** is very thick, 2 cm. or more in thickness, fleshy, soft, very juicy, and in wet weather very clammy and somewhat sticky to the touch. When mature there are lines of color of different shades extending out radially on the upper surface, and in making a longitudinal section of the cap there are quite prominent, alternating, dark and light red lines present in the flesh. The **tubes**, short at first, become 2–3 mm. long, they are yellowish or tinged with flesh color, becoming soiled in age. The **spores** are elliptical, yellowish, and 5–6 μ long.

The plant occurs on dead trunks or stumps of oak, chestnut, etc., in wet weather from June to September. I have usually found it on chestnut.

The beef-steak fungus is highly recommended by some, while others are not pleased with it as an article of food. It has an acid flavor which is disagreeable to some, but this is more marked in young specimens and in those not well cooked. When it is sliced thin and well broiled or fried, the acid taste is not marked.

- PLATE 65.
- Fig. 1. — Fistulina hepatica.
- Fig. 2. — F. pallida.
- Copyright 1900.

Fistulina pallida B. & Rav. (*Fistulina firma* Pk.) — This rare and interesting species was collected by Mrs. A. M. Hadley, near Manchester, New Hampshire, October, 1898, and was described by Dr. Peck in the Bulletin of the Torrey Botanical Club, **26**: 70, 1899, as [Pg 187] *Fistulina firma*. But two plants were then found, and these were connected at the base. During August and September it was quite common in a small woods near Ithaca, N. Y., and was first collected growing from the roots of a dead oak stump, August 4 (No. 3227 C. U. herbarium), and afterward during October. During September I collected it at Blowing Rock, N. C., in the Blue Ridge mountains, at an elevation of nearly 5000 feet, growing from the roots of a dead white oak tree. It was collected during September, 1899, by Mr. Frank Rathbun at Auburn, N. Y. It was collected by Ravenel in the mountains of South Carolina, around a white oak stump by Peters in Alabama, and was first described by Berkeley in 1872, in **Grev. 1**: 71, Notices of N. A. F. No. 173. Growing from roots or wood underneath the surface of the ground, the plant has an erect stem, the length of the stem depending on the depth at which the root is buried, just as in the case of *Polyporus radicatus*, which has a similar habitat. The plants are 5–12 cm. high, the cap is 3–7 cm. broad, and the stem 6–8 mm. in thickness.

Plate 66, Figure 180. — Fistulina pallida. Cap wood-brown to fawn or clay color, tubes and lower part of the stem whitish (natural size). Copyright.

The **pileus** is wood brown to fawn, clay color or isabelline color. It is nearly semi-circular to reniform in outline, and the margin broadly crenate, or sometimes lobed. The stem is attached at the concave margin, where the cap is auriculate and has a prominent boss or elevation, and bent at right angles with a characteristic cur-ve. The pileus is firm, flexible, tough and fibrous, flesh white. The

302

surface is covered with a fine and dense tomentum. The pileus is 5–8 mm. thick at the base, thinning out toward the margin. The **tubes** are whitish, 2–3 mm. long and 5–6 in the space of a millimeter. They are very slender, tubular, the mouth somewhat enlarged, the margin of the tubes pale cream color and minutely mealy or furfuraceous, with numerous irregular, roughened threads. The tubes often stand somewhat separated, areas being undeveloped or younger, so that the surface of the under side is not regular. The tubes are not so crowded as is usual in the *Fistulina hepatica*. They are not decurrent, but end abruptly near the stem. The **spores** are subglobose, 3 μ in diameter. The stem tapers downward, is whitish below, and near the pileus the color changes rather abruptly to the same tint as the pileus. The stem is sometimes branched, and two or three caps present, or the caps themselves may be joined, as well as the stems, so that occasionally very irregular forms are developed, but there is always the peculiar character of the attachment of the stem to the side of the cap.

Figure 180 is from plants (No 3676, C. U. herbarium) collected [Pg 188] at Blowing Rock, N. C., September, 1899. Figures on the colored plate represent this plant.

Polyporus frondosus Fr. **Edible.**—This plant occurs in both Europe and America, and while not very common seems to be widely distributed. It grows about old stumps or dead trees, from roots, often arising from the roots below the surface of the ground, and also is found on logs. The plant represents a section of the genus *Polyporus*, in which the body, both the stem and the cap, are very much branched. In this species the stem is stout at the base, but it branches into numerous smaller trunks, which continue to branch until finally the branches terminate in the expanded and leaf-like caps as shown in Figs. 181–182. The plants appear usually during late summer and in the autumn. The species is often found about oak stumps. Some of the specimens are very large, and weigh 10 to 20 pounds, and the mass is sometimes 30 to 60 cm. (1–2 feet) in diameter.

The plant, when young and growing, is quite soft and tender, though it is quite firm. It never becomes very hard, as many of the other species of this family. When mature, insects begin to attack it, and not being tough it soon succumbs to the ravages of insects and decay, as do a number of the softer species of the *Polyporaceæ*. The caps are very irregular in shape, curved, repand, radiately furrowed, sometimes zoned; gray, or hair-brown in color, with a perceptibly hairy surface, the hairs running in lines on the surface. Sometimes they are quite broad and not so numerous as in Plate 67, and in other plants they are narrow and more numerous, as in Plate 68. The tubes are more or less irregular, whitish, with a yellowish tinge when old. From the under side of the cap they extend down on the stem. When the spores are mature they are sometimes so numerous that they cover the lower caps and the grass for quite a distance around as if with a white powder.

This species is edible, and because of the large size which it often attains, the few plants which are usually found make up in quantity what they lack in numbers. Since the plant is quite firm it will keep several days after being picked, in a cool place, and will serve for several meals. A specimen which I gathered was divided between two families, and was served at several meals on successive days. When stewed the plant has for me a rather objectionable taste, but the stewing makes the substance more tender, and when this is followed by broiling or frying the objectionable taste is removed and it is quite palatable. The plants represented in Plates 67 and 68 were collected at Ithaca.

Plate 67, Figure 181.—Polyporus frondosus. Caps hair-brown or grayish, tubes white (1/3 natural size, masses often 20–40 cm. in breadth). The caps in this specimen are quite broad, often they are narrower as in Fig. 182. Copyright.

[Pg 189]

Plate 68, Figure 182.—Polyporus frondosus. Side and under view of a larger cluster (1/3 natural size). Copyright.

There are several species which are related to the frondose poly-porus which occur in this country as well as in Europe. **Polyporus intybaceus** Fr., is of about the same size, and the branching, and form of the caps is much the same, but it is of a yellowish brown or reddish brown color. It grows on logs, stumps, etc., and is probably edible. It is not so common at Ithaca as the frondose polyporus.

Figure 183.—Polyporus umbellatus. Caps hair-brown (natural si-ze, often much larger). Copyright.

Polyporus umbellatus Fr.—This species is also related to the frondose polyporus, but is very distinct. It is more erect, the bran-ching more open, and the caps at the ends of the branches are more or less circular and umbilicate. The branches are long, cylindrical and united near the base. The spreading habit of the branching, or the form of the caps, suggests an umbel or umbrella, and hence the specific name *umbellatus*.

The tufts occur from 12–20 cm. in diameter, and the individual caps are from 1–4 cm. in diameter. It grows from underground roots and about stumps during summer. It is probably edible, but I have

[Pg 190] never tried it. Figure 183 is from a plant (No. 1930, C. U. herbarium) collected in Cascadilla woods, Ithaca.

Polyporus sulphureus (Bull.) Fr. **Edible.** (*Boletus caudicinus* Schaeff. T. 131, 132: *Polyporus caudicinus* Schroeter, Cohn's Krypt. Flora, Schlesien, p. 471, 1899).—The sulphur polyporus is so-called because of the bright sulphur color of the entire plant. It is one of the widely distributed species, and grows on dead oak, birch, and other trunks, and is also often found growing from wounds or knotholes of living trees of the oak, apple, walnut, etc. The mycelium enters at wounds where limbs are broken off, and grows for years in the heart wood, disorganizing it and causing it to decay. In time the mycelium has spread over a considerable area, from which nutriment enough is supplied for the formation of the fruiting condition. The caps then appear from an open wound when such an exit is present.

The color of the plant is quite constant, but varies of course in shades of yellow to some extent. In form, however, it varies greatly. The caps are usually clustered and imbricated, that is, they overlap. They may all arise separately from the wood, and yet be overlapping, though oftener several of them are closely joined or united at the base, so that the mass of caps arises from a common outgrowth from the wood as shown in Fig. 184. The individual caps are flattened, elongate, and more or less fan-shaped. When mature there are radiating furrows and ridges which often increase the fan-like appearance of the upper surface of the cap. Sometimes also there are more or less marked concentric furrows. The caps may be convex, or the margin may be more or less upturned so that the central portion is depressed. When young the margin is thick and blunt and of course lighter in color, but as the plant matures the edge is usually thinner.

In some forms of the plant the caps are so closely united as to form a large rounded or tubercular mass, only the blunt tips of the individual caps being free. This is well represented in Fig. 185, from a photograph of a large specimen growing from a wound in a butter-nut tree in Central New York. The plant was 30 cm. in diameter. The plants represented in Plate 69 grew on an oak stump. The tree was affected by the fungus while it was alive, and the heart wood became so weakened that the tree broke, and later the fruit form of the fungus appeared from the dead stump.

Plate 69, Figure 184.—Polyporus sulphureus, on oak stump. Entirely sulphur-yellow (1/6 natural size). Copyright.

The tubes are small, and the walls thin and delicate, and are sometimes much torn, lacerated, and irregular. When the mycelium has grown in the interior of a log for a number of years it tends to grow in sheets along the line of the medullary rays of the wood or [Pg 191] across in concentric layers corresponding to the summer wood. Also as the wood becomes more decomposed, cracks and rifts appear along these same lines. The mycelium then grows in abundance in these rifts and forms broad and extensive sheets which resemble somewhat chamois skin and is called "punk." Similar punk is sometimes formed in conifers from the mycelium of *Fomes pinicola*.

Plate 70, Figure 185.—Polyporus sulphureus. Caps joined in a massive tubercle (1/2 natural size).

Polyporus sulphureus has long been known as an edible fungus, but from its rather firm and fibrous texture it requires a different preparation from the fleshy fungi to prepare it for the table, and this may be one reason why it is not employed more frequently as an article of food. It is common enough during the summer and especially during the autumn to provide this kind of food in considerable quantities.

Plate 71, Figure 186.—Polyporus brumalis. Cap and stem brown, tubes white. Lower three plants natural size, upper one enlarged twice natural size. Copyright.

Polyporus brumalis (Pers.) Fr.—This pretty plant is found at all seasons of the year, and from its frequency during the winter was named *brumalis*, from *bruma*, which means winter. It grows on sticks and branches, or on trunks. It usually occurs singly, sometimes two or three close together. The plants are 3–6 cm. high, the cap 2–6 cm. in diameter, and the stem is 3–6 mm. in thickness.

The **cap** is convex, then plane, and sometimes depressed at the center or umbilicate. When young it is somewhat fleshy and pliant, then it becomes tough, coriaceous, and hard when dry. During wet weather it becomes pliant again. Being hard and firm, and tough, it preserves long after mature, so that it may be found at any season of the year. The cap is smoky in color, varying in shade, sometimes very dark, almost black, and other specimens being quite light in color. The surface is hairy and the margin is often fimbriate with coarse hairs. The **stem** is lighter, hairy or strigose. The **tubes** are first white, then become yellowish. The tubes are very regular in arrangement.

Figure 186 represents well this species, three plants being grouped rather closely on the same stick; two show the under surface and one gives a side view. The upper portion of the plate represents two of the plants enlarged, the three lower ones being natural size. The plant is very common and widely distributed over the world. Those illustrated in the plate were collected at Ithaca. This species is too tough for food.

Many of the thin and pliant species of *Polyporus* are separated by some into the genus **Polystictus**. The species are very numerous, as well as some of the individuals of certain species. They grow on wood or on the ground, some have a central stem, and others are shelving, while some are spread out on the surface of the wood. One [Pg 192] very pretty species is the **Polystictus perennis** Fr. This grows on the ground and has a central stem. The plant is 2–3 cm. high, and the cap 1–4 cm. broad. The **pileus** is thin, pliant when fresh and somewhat brittle when dry. It is minutely velvety on the upper surface, reddish brown or cinnamon in color, expanded or umbilicate to nearly funnel-shaped. The surface is marked beautifully by radiations and fine concentric zones. The **stem** is also velvety. The **tubes** are minute, the walls thin and acute, and the mouths angular and at last more or less torn. The margin of the cap is finely fimbriate, but in old specimens these hairs are apt to become rubbed off. The left hand plant in Fig. 187 is *Polyporus perennis*.

Polystictus cinnamomeus (Jacq.) Sacc., (*P. oblectans* Berk. Hook. Jour. p. 51, 1845, Dec. N. A. F. No. 35: *P. splendens* Pk., 26th Report N. Y. State Mus., p. 26) is a closely related species with the same habit, color, and often is found growing side by side with *P. perennis*. The margin of the cap is deeply and beautifully lacerate, as shown in the three other plants in Fig. 187. *Polystictus connatus* Schw., grows in similar situations and one sometimes finds all three of these plants near each other on the ground by roadsides. *P. connatus* has much larger pores than either of the other two, and it is a somewhat larger plant. Figure 187 is from a photograph of plants collected at Blowing Rock, N. C., during September, 1899.

Figure 187.—Left-hand plant Polystictus perennis; right-hand three plants Polystictus cinnamomeus. All natural size. Copyright.

Polystictus versicolor (L.) Fr., is a very common plant growing on trunks and branches. It is more or less shelving, with a leaf-like pileus, marked by concentric bands of different colors. **P. hirsutus** Fr., is a somewhat thicker and more spongy plant, whitish or grayish in color, with the upper surface tomentose with coarse hairs. **P. cinnabarinus** (Jacq.) Fr., is shelving, spongy, pliant, rather thick, cinnabar colored. It grows on dead logs and branches. It is sometimes [Pg 193] placed in the genus *Trametes* under the same specific name. **Polystictus pergamenus** Fr., is another common one growing on wood of various trees. It is thin and very pliant when fresh, somewhat tomentose above when young, with faint bands, and the tubes are often violet or purple color, and they soon become deeply torn and lacerate so that they resemble the teeth of certain of the hedgehog fungi.

Plate 72, Figure 188.—Polyporus lucidus. Caps bright red or chestnut color, with a hard shiny crust (1/6 natural size). Copyright.

Polyporus lucidus (Leys.) Fr. [*Fomes lucidus* (Leys.) Fr.]—This species is a very striking one because of the bright red or chestnut color, the hard and brittle crust over the surface of the cap, which has usually the appearance of having been varnished. It grows on trunks, logs, stumps, etc., in woods or groves. The cap is 5–20 cm. in diameter, and the stem is 5–20 cm. long, and 1–2 cm. in thickness. The stem is attached to one side of the pileus so that the pileus is lateral, though the stem is more or less ascending.

The **cap** is first yellowish when young, then it becomes blood red, then chestnut color. The **stem** is the same color, and the **tubes** are not so bright in color, being a dull brown. The substance of the plant is quite woody and tough when mature. When dry it is soon attacked and eaten by certain insects, which are fond of a number of fungi, so that they are difficult to preserve in good condition in herbaria without great care.

The surface of the pileus is quite uneven, wrinkled, and coarsely grooved, the margin sometimes crenate, especially in large specimens. Figure 188 represents the plant growing on a large hemlock spruce stump in the woods. The surface character of the caps and the general form can be seen. This photograph was taken near Ithaca, N. Y.

Polyporus applanatus (Pers.) Fr. [*Fomes applanatus* (Pers.) Wallr.]—This plant is also one of the very common woody *Polyporaceæ*. It grows on dead trunks, etc., and sometimes is found growing from the wounds of living trees. It is very hard and woody. It has a hard crust, much harder than that of the *Polyporus lucidus*. The surface is more or less marked by concentric zones which mark off the different years' growth, for this plant is perennial. At certain seasons of the year the upper surface is covered with a powdery substance of a reddish brown color, made up of numerous colored spores or conidia which are developed on the upper surface of this plant in addition to the smaller spores developed in the tubes on the under surface.

The plant varies in size from 5–20 cm. or more in diameter, and 1–10 cm. in thickness, according to the rapidity of growth and the age of the fungus. The fruiting surface is white, and the tubes are very [Pg 194] minute. They scarcely can be seen with the unaided eye. Bruises of the tubes turn brown, and certain "artists" often collect these plants and sketch with a pointed instrument on the tube surface. For other peculiarities of this plant see page 15. The age of the plant can usually be told by counting the number of the broader zones on the upper surface, or by making a section through the plant and counting the number of tube strata on the lower surface of the cap at its base.

Polyporus leucophæus Mont., is said to differ from this species in being more strongly zonate, and in the crust being whitish instead of reddish brown.

Polyporus fomentarius (L.) Fr. [*Fomes fomentarius* (L.) Fr.,] is hoof-shaped, smoky in color, or gray, and of various shades of dull brown. It is strongly zoned and sulcate, marking off each year's growth. The margin is thick and blunt, and the tube surface concave, the tubes having quite large mouths so that they can be readily seen, the color when mature being reddish brown. Sections of the plant show that the tubes are very long, the different years' growth not being marked off so distinctly as in *P. applanatus* and *leucophæus*. The plant grows on birch, beech, maple, etc. The inner portion was once used as tinder.

Polyporus pinicola (Swartz.) Fr. [*Fomes pinicola* (Swartz.) Fr.] occurs on dead pine, spruce, balsam, hemlock spruce, and other conifers. The cap is about the width of the *F. applanatus*, but it is stouter, and does not have the same hard crust. The young growth at the margin, which is very thick, is whitish yellow, while the old zones are reddish. The tubes are yellowish, and sections show that they are in strata corresponding to the years' growth. **Polyporus igniarius** (L.) Fr. [*Fomes igniarius* (L.) Fr.] is a black species, more or less triangular, or sometimes hoof-shaped. The yearly zones are smaller, become much cracked, and the tubes are dark brown. One of these plants which I found on a birch tree in the Adirondacks was over 80 years old.

The genus *Merulius* has a fruiting surface of irregular folds or wrinkles, forming shallow, irregular pits instead of a deeply honey-combed surface. **Merulius lacrymans** (Jacq.) Fr., the "weeping" merulius, or "house fungus," often occurs in damp cellars, buildings, conduit pipes, etc. It is very destructive to buildings in certain parts of Europe (see Figs. 189, 190). **Merulius tremellosus** Schrad., is very common in woods during autumn. It is of a gelatinous consistency, and spread on the under surface of limbs or forms irregular shelves from the side (see Figs. 191, 192).

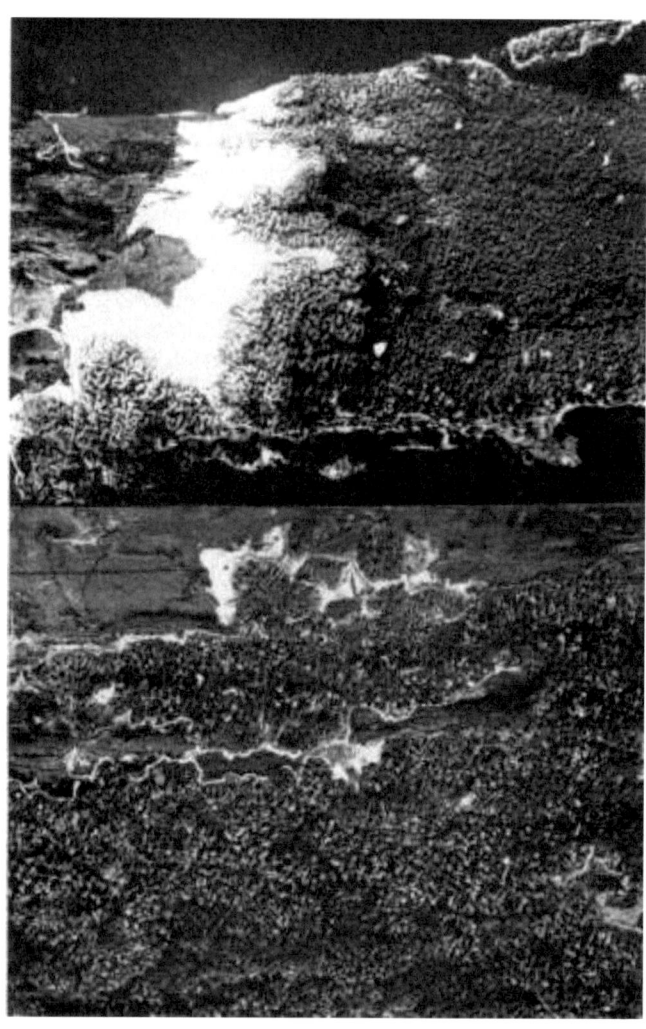

Plate 73.—Merulius lacrymans. Figure 189.—Upper plant in conduit pipe leading from wash room, Gymnasium C. U., Autumn, 1899. Figure 190.—Lower plant from under surface decaying hemlock spruce log in woods near Freeville, N. Y., October, 1899. Margin of plants white, fruiting surface a network of irregular folds, golden brown, or brown. Copyright.

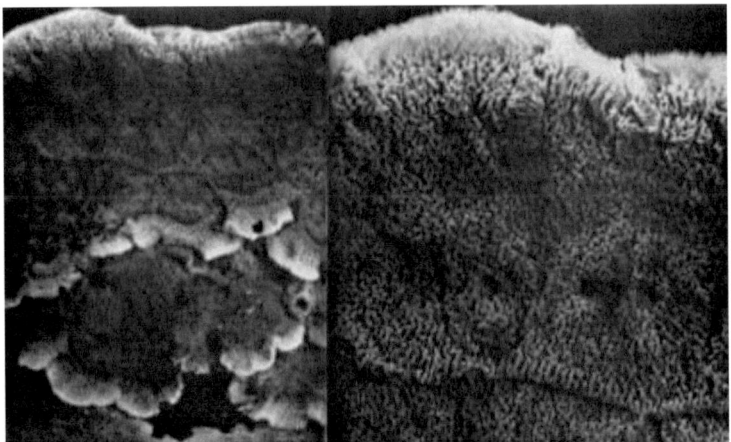

Plate 74. — Merulius tremellosus. Figure 191. — Natural size.

Figure 192. — Enlarged to show character of fruiting surface. Fruiting surface yellowish; margin and upper surface in shelving forms, white, hairy. Copyright.

Plate 75, Figure 193.—Phlebia merismoides. On rotting log, woods near Ithaca, November 23, 1898 (No. 2634 C. U. herbarium). Various shades of orange, yellow or yellow brown when old. Copyright.

Plate 76, Figure 194.—Phlebia merismoides. Portion of a plant 2-1/2 times natural size, to show interrupted folds of fruiting surface. For colors see Fig. 193. Copyright.

CHAPTER X. [Pg 195]

HEDGEHOG FUNGI: HYDNACEAE.

The plants belonging to this family vary greatly in size, form, and consistency. Some of them are very large, some quite small, some are fleshy in consistency, some are woody, corky; some membranaceous; and if we include plants formerly classed here, some are gelatinous, though there is a tendency in recent years on the part of some to place the gelatinous ones among the trembling fungi. The special character which marks the members of this family is the peculiarity of the fruiting surface, just as a number of the other families are distinguished by some peculiarity of the fruiting surface. In the *Hydnaceæ* it covers the surface of numerous processes in the form of spines, teeth, warts, coarse granules, or folds which are interrupted at short intervals. These spines or teeth always are directed toward the earth when the plant is in the position in which it grew. In this way the members of the family can be distinguished from certain members of the club fungi belonging to the family *Clavariaceæ*, for in the latter the branches or free parts of the plant are erect.

In form the *Hydnaceæ* are shelving, growing on trees; or growing on the ground they often have a central or eccentric stem, and a more or less circular cap; some of them are rounded masses, growing from trees, with very long spines extending downward; others have ascending branches from which the spines depend; and still others form thin sheets which are spread over the surface of logs and sticks, the spines hanging down from the surface, or roughened with granules or warts, or interrupted folds (see *Phlebia*, Figs. 193, 194). In one genus there is no fruit body, but the spines themselves extend downward from the rotten wood, the genus *Mucronella*. This is only distinguished, so far as its family position is concerned, from such a species as *Clavaria mucida* by the fact that the plant grows downward from the wood, while in *C. mucida* it grows erect.

HYDNUM Linn.

The only species of the *Hydnaceæ* described here are in the genus *Hydnum*. In this genus the fruiting surface is on spine, or awl-shaped processes, which are either simple or in some cases the tips are more or less branched. The plants grow on the ground or on [Pg 196] wood. The species vary greatly in form. Some are provided with a more or less regular cap and a stem, while others are shelving or bracket shaped, and still others are spread out over the surface of the wood (resupinate).

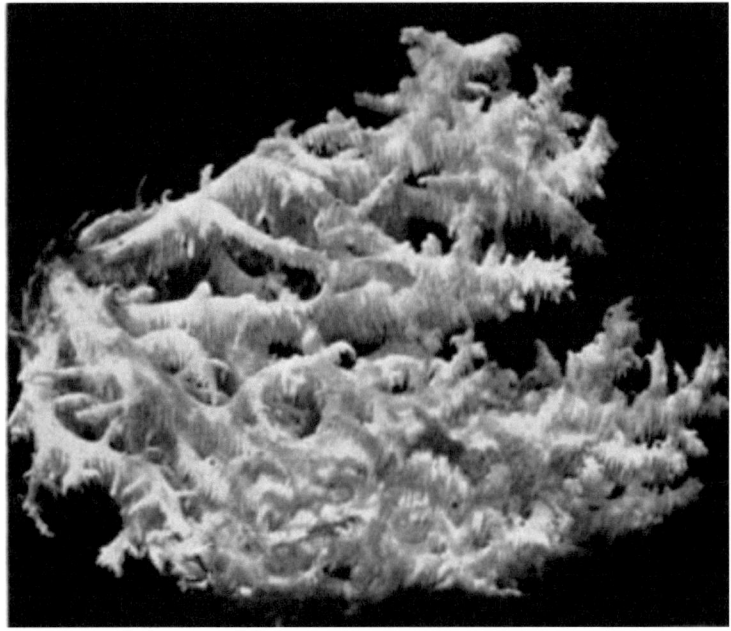

Figure 195.—Hydnum coralloides. Entirely white (natural size). Copyright.

Hydnum coralloides Scop. **Edible.**—Among the very beautiful species of the genus *Hydnum* is the coral one, *Hydnum coralloides*. It grows in woods forming large, beautiful, pure white tufts on rotten logs, branches, etc. The appearance of one of these tufts is shown in Fig. 195. There is a common stem which arises from the wood, and this branches successively into long, ascending, graceful shoots. The spines are scattered over the entire under side of these branches and hang down for 3-6 mm. They are not clustered at the ends of the branches, as in the bear's head hydnum, and the species can be easily distinguished by giving attention to the form of the branching and the distribution of the spines on the under side of the branches. Figure 195 represents a plant collected at Ithaca, and it is natural [Pg 197] size. They grow, however, much larger than this specimen. The species is widely distributed, and not uncommon. It is excellent for food.

Plate 77, Figure 196.—Hydnum caput-ursi. Entirely white (natural size). Copyright.

Hydnum caput-ursi Fr. **Edible.**—This plant is also a beautiful one. It is more common than the coral hydnum so far as my observation goes. It is known by the popular name of "bear's head hydnum" in allusion to the groups of spines at the ends of the branches. It occurs in woods with a similar habit of growing on trunks, branches, etc. This plant also arises from the wood with a single stout stem, which then branches successively, the ends of the branches having groups of long pendant spines appearing like numerous heads. Sometimes the spines on the top of the group are twisted or curled in a peculiar way. Large tufts are sometimes formed, varying from 12–20 or more centimeters in diameter. Figure 196 is from a plant collected at Ithaca.

[Pg 198]

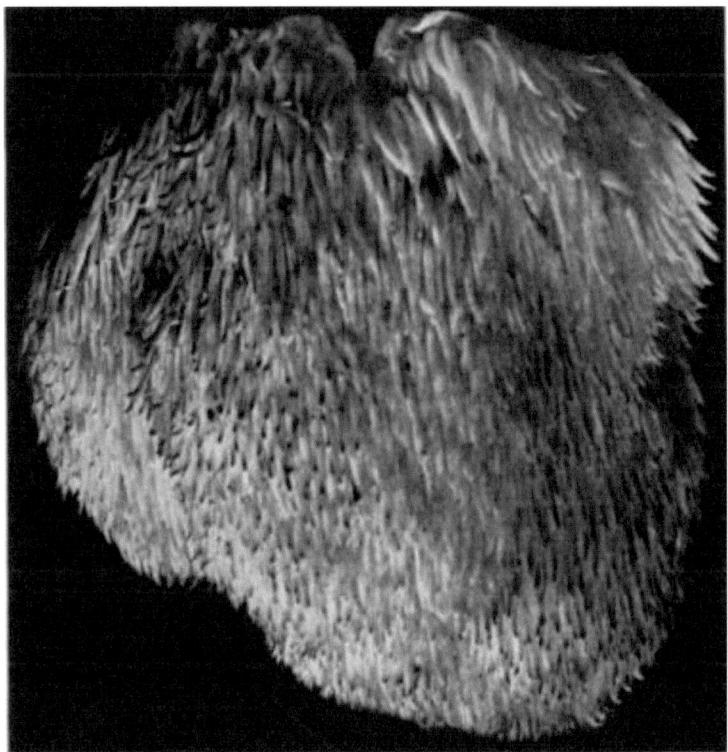

Figure 197.—Hydnum erinaceus. Entirely white (natural size, often larger).

Hydnum caput-medusæ Bull. **Edible.**—The medusa's head hydnum is a rarer species than either of the above in this country. It forms a large, tubercular mass which does not branch like the coral hydnum or the bear's head, but more like the Satyr's beard hydnum, though the character of the spines will easily separate it from the latter. The spines cover a large part of this large tubercle, and hang downward. The plant is known by the additional character, that, on the upper part of the tubercle, the spines are twisted and interwoven in a peculiar fashion.

Hydnum erinaceus Bull. **Edible.** — This plant is sometimes called "Satyr's beard." It grows on dead trunks in the woods or groves, and is often found growing from wounds in living trees. It forms a large, tubercular mass which does not branch. The spines are very long and straight and hang downward in straight parallel lines from the sides of the mass. The spines are from 1–2 cm. or more long. Figure 197 represents one of the plants, showing the long spines.

Hydnum repandum L. **Edible.** — This plant is not uncommon, and it is widely distributed. It grows usually in woods, on the ground. It varies greatly in size, from very small specimens, 1–2 cm. high to others 10–12 cm. high. The cap is 2–18 cm. broad, and the stem 6–12 mm. in thickness.

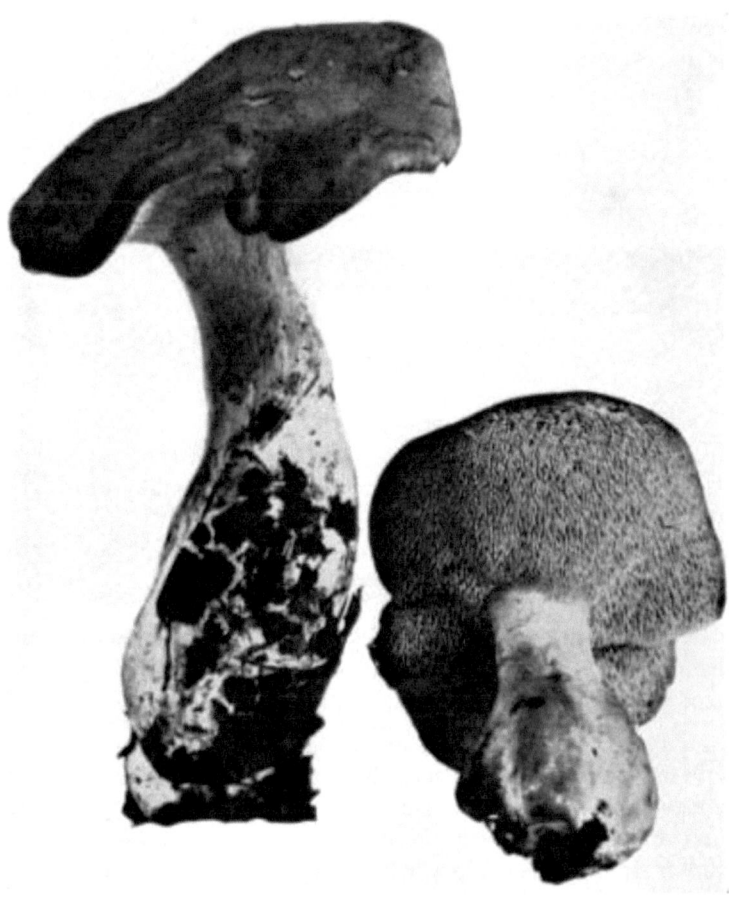

Plate 78, Figure 198. — Hydnum repandum. Cap whitish or yellowish, or pale yellowish brown; spines whitish or yellowish (natural size, often smaller). Copyright.

It is entirely white or the cap varies to buff, dull yellow reddish or dull brown. It is very brittle, and must be handled with the utmost care if one wishes to preserve the specimen intact. The pileus is more or less irregular, the stem being generally eccentric, so that the pileus is produced more on one side than on the other, sometimes entirely lateral at the end of the stem. The margin is more or less wavy or repand. The spines are white, straight, and very brittle. The stem is even or clavate. Figure 198 is from plants collected at Ithaca during August, 1899, and represents one of the large specimens of the species. In one plant the pileus is entirely lateral on the end of the long clavate stem, and is somewhat reniform, the stem being attached at the sinus. In the other plant the stem is attached near the center. This species is considered one of the best mushrooms for the table.

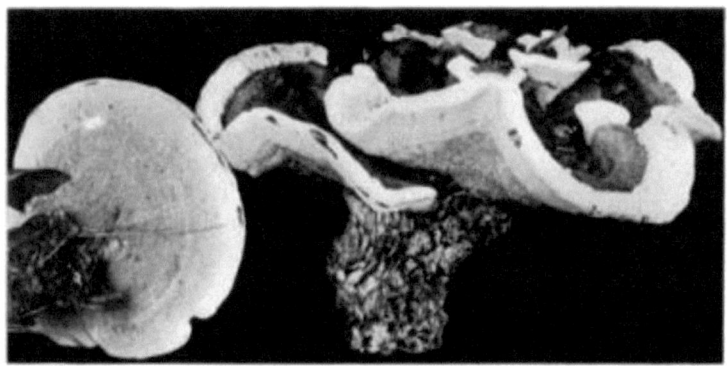

Plate 79, Figure 199. — Hydnum putidum. Caps whitish then buff, then brownish or nearly black in older parts, edge white (natural size). Copyright.

Hydnum imbricatum L. **Edible.** — This is a very variable species both in size and in the surface characters of the pileus. It occurs in woods, groves, or in open places under trees. The plants are 3–7 cm. high, and the pileus varies from 5–15 cm. broad, the stem from .5–2.5 cm. in thickness. The pileus is convex and nearly expanded, fleshy, thinner at the margin, regular or very irregular. The color is grayish in the younger and smaller plants to umber or quite dark in the [Pg 199] larger and older ones. The surface is cracked and torn into triangular scales, showing the whitish color of the flesh between the scales. The scales are small in the younger plants and larger in the older ones. Figure 200 is from plants collected at Ithaca, and the pileus in these specimens is irregular. The species is edible, but bitter to the taste.

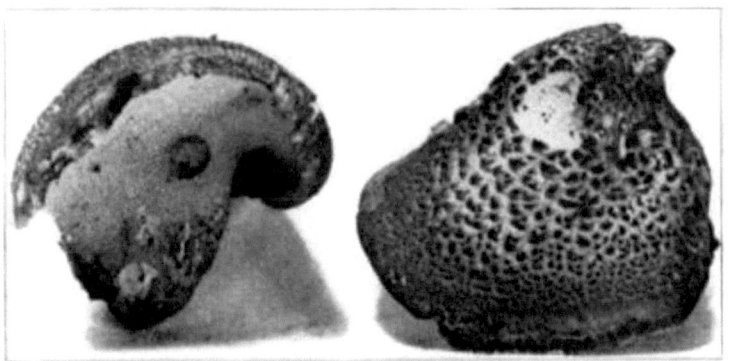

Figure 200.—Hydnum imbricatum. Caps brownish, spines whitish (natural size, often larger).

Hydnum putidum Atkinson.—This plant grows on the ground in woods, and was collected in the Blue Ridge mountains at Blowing Rock, N. C., at an elevation of about 4000 feet. It is remarkable for its peculiar odor, resembling, when fresh, that of an Ethiopian; for its tough, zonate pileus with a prominent white edge, and the stout irregular stem, resembling the stem of *Hydnum velutinum*. The plants are 8–12 cm. high, the cap 8–12 cm. broad, and the stem 2–4 cm. in thickness. The plants grow singly, or sometimes a few close together, and then two or more may be conjoined.

The **pileus** is first umbilicate or depressed, becoming depressed or infundibuliform, irregular, eccentric, the margin repand, and sometimes lobed, and lobes appearing at times on the upper surface of the cap. The surface is first tomentose or pubescent, becoming smooth, with prominent concentric zones probably marked off by periodical growth; the color is first white, so that the edge is white, becoming cream color to buff, and in age dull brown and sometimes blackish brown in the center of the old plants. The pubescence disappears from the old portions of the cap, so that it is smooth. The pubescence or tomentum is more prominent on the intermediate zones. The margin is rather thick, somewhat acute or blunt, the upper portion [Pg 200] of the flesh is spongy and the middle portion tough and coriaceous, and darker in color. The pileus is somewhat pliant when moist or wet, and firm when dry, the dark inner stratum hard.

The **spines** are first white or cream color, in age changing through salmon color, or directly into grayish or grayish brown. The spines when mature are long, slender, crowded, and decurrent on the upper part of the stem. The **spores** are white, globose, echinulate, 3–4 μ. The **stem** is stout and irregular, very closely resembling the stem of *Hydnum velutinum*, with a thick, spongy, outer layer and a central hard core.

The odor, which resembles that of a perspiring darkey, before the plant is dry, disappears after drying, and then the plant has the same agreeable odor presented by several different species of Hydnum. The odor suggests *H. graveolens*, but the characters of the stem and surface of the pileus separate it from that species, while the tough and pliant character of the cap separates it from *H. fragile*. Figure 199 is from plants (No. 4334, C. U. herbarium) collected at Blowing Rock, N. C., during September, 1899.

CHAPTER XI.

CORAL FUNGI: CLAVARIACEAE.

This family is a very characteristic one, and very interesting from the large number of beautiful species in one genus, the genus *Clavaria*. The plants all are more or less erect, or at least stand out from the substratum, that is, the substance on which they are growing. The fruiting surface covers the entire upper part of the plant, all but the bases of the stems. Some of the branched species of the *Thelephoraceæ* resemble the branched species of the *Clavariaceæ*, but in the former there is a more or less well defined upper portion on the tips of the branches which is flat, or truncate, and sterile, that is, lacks the fruiting surface. Some of the species are simple, elongate and clavate bodies. Some stand singly, others are clustered, or others are joined by their bases, and others still are very much branched. All of the species are said to be edible, that is, they are not poisonous. A few are rather tough, but they are mostly the small species which would not be thought of for food. The spores are borne on club-shaped basidia, as in the common mushrooms.

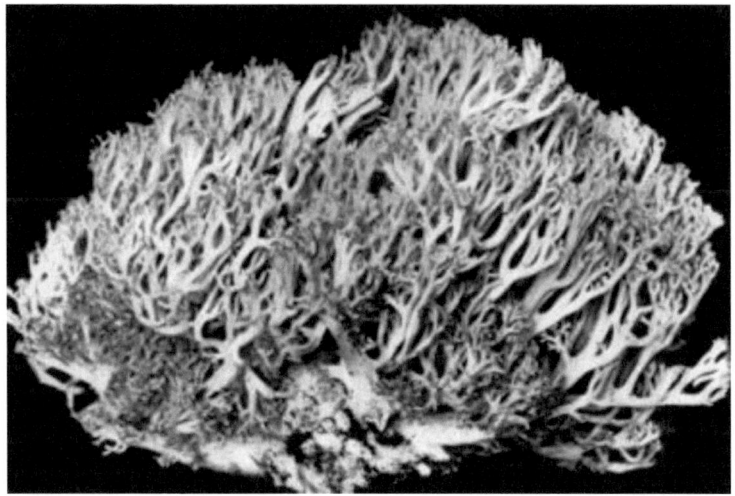

Plate 80, Figure 201.—Clavaria formosa. Yellowish, red tipped when young, red disappearing in age (natural size, sometimes twice this size). Copyright.

CLAVARIA Vaill. [Pg 201]

The genus *Clavaria* is one of the most common ones in the family, and is one of the most attractive from the variety and beauty of several of the species. All of the plants are more or less erect, and at least stand out from the substratum on which they grow. They are either long and simple and more or less club-shaped, as the name implies, or they are branched, some but a few times, while others are very profusely branched. The plants vary in color, some are white, some yellow, some red, and some are red-tipped, while others are brownish in color.

Figure 202. — Clavaria botrytes. Branches red tipped (natural size).

Clavaria formosa Pers. **Edible.** — This is one of the handsomest of the genus. It is found in different parts of the world, and has been collected in New England and in the Carolinas in this country. It is usually from 15–20 cm. high, and because of the great number of branches is often broader in extent. There is a stout stem from 2–4 cm. in diameter, deep in the ground. This branches into a few stout trunks, which then rapidly branch into slender and longer branches, [Pg 202] terminating into numerous tips. The entire plant is very brittle, and great care is necessary to prevent its breaking, both before drying and afterward. When the plant is young and is just pushing out of the ground, the branches, especially the tips, are bright colored, red, pink, or orange, the color usually brighter when young in the younger plants. As the plant becomes older the color fades out, until at maturity the pink or red color has in many cases disappeared, and then the entire plant is of a light yellowish, or of a cream buff color. The spores are in mass light yellow, and the spores on the surface of the plant probably give the color to the plant at this stage. The spores are long, oval or oblong, 10–15 × 2.5–3 μ, and are minutely spiny. Figure 201 is from a plant (No. 4343, C. U. herbarium) collected at Blowing Rock, N. C., in September, 1899. The plant is very common in the mountain woods of North Carolina.

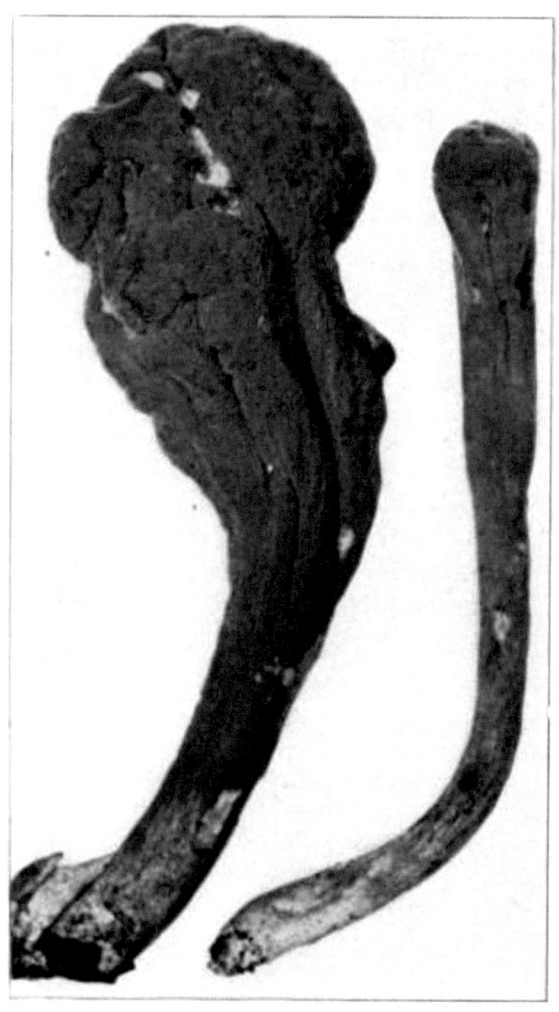

Figure 203.—Clavaria pistillaris. Dull whitish, tan or reddish (natural size).

Specimens of this Clavaria were several times prepared for table use during my stay in the mountains, but the flavor was not an agreeable one, possibly due to the fact that it needs some special preparation and seasoning.

Clavaria botrytes Pers. **Edible.**—This plant is much smaller than *C. formosa*, but has much the same general habit and color, especially when *C. formosa* is young. The plant has a stout stem which soon [Pg 203] dissolves into numerous branches, which are red tipped. The spores are white, and in this way it may be distinguished from *C. formosa*, or from *Clavaria aurea* (Schaeff.), which has yellow or ochre spores, and which has also much the same habit as *C. botrytes*, and is nearer in size.

Figure 204.—Clavaria mucida. White (natural size). Copyright.

Clavaria pistillaris Linn. **Edible.**—This plant is a characteristic one because of its usually large size and simple form. It is merely a club-shaped body, growing from the ground. It has a wide range, both in Europe and North America, but does not seem to be common, though I have found it more common in the mountain woods of North Carolina than in New York. The plant is 5–20 cm. high, and 1–3 cm. thick at the upper end. It is smooth, though often irregularly grooved and furrowed, due probably to unequal tensions in growth. The apex in typical specimens is rounded and blunt. It is dull white or tan color or rufescent. The flesh is white, and very spongy, especially in age, when it is apt to be irregularly fistulose. Figure 203 is from plants collected at Blowing Rock, N. C., during September 1899.

There is what seems to be an abnormal form of this species figured by Schaeffer, Table 290, which Fries separated as a distinct species and placed in the genus *Craterellus*, one of the *Thelephoraceæ*, and called by him *Craterellus pistillaris*. This plant has been found at Ithaca, and the only difference between this and the *Clavaria pistilla-*

ris L., seems to be in the fact that in *Craterellus pistillaris* the end is truncate or in some specimens more or less concave. The spores seem to be the same, and the color and general habit of the two plants are the same. It is probably only a form of *Clavaria pistillaris.*

Clavaria mucida Pers. — This is one of the smallest species of the genus *Clavaria*. It grows on rotten wood, and appears throughout [Pg 204] the year. It is usually simple and clavate, but sometimes branched. The plant is white, or yellowish, or sometimes rose color, and measures from 0.5 to 2 cm. in height, though I have usually found it from 0.5–1 cm. in height. It is soft and watery. Figure 204 is from plants (No. 4998, C. U. herbarium) collected at Ithaca in October, 1899.

CHAPTER XII.

THE TREMBLING FUNGI: TREMELLINEAE.

These fungi are called the trembling fungi because of their gelatinous consistency. The colors vary from white, yellow, orange, reddish, brownish, etc., and the form is various, often very irregular, leaf-like, or strongly folded and uneven. They are when fresh usually very soft, clammy to the touch, and yielding like a mass of gelatine. They usually grow on wood, but some species grow on the ground, and some are parasitic. The fruit surface usually covers the entire outer surface of the plant, but in some it is confined to one side of the plant. The basidia are peculiar to the order, are deeply seated in the substance of the plant, rounded or globose, and divided into four cells in a cruciate manner. From each one of these cells of the basidium a long, slender process (sterigma) grows out to the surface of the plant and bears the spore. A few species only are treated of here.

TREMELLA Dill.

In this genus the plants are gelatinous or cartilaginous. The form of the plant is usually very much contorted, fold-like or leaf-like, and very much branched. The fruiting surface extends over the entire upper surface of the plant.

[Pg 205]

Figure 205.—Tremella mycetophila, on Collybia dryophila (natural size).

Tremella lutescens Pers.—This plant is entirely yellow, and occurs on branches. It is 2–5 cm. in diameter, and is strongly folded, somewhat like the folds of a brain (gyrose). It is very soft and inclined to be watery and fluid, and is of a bright yellow color, spread out on the surface of rotten wood. It is of world-wide distribution, and appears from mid-summer to late autumn.

Tremella mycetophila Pk.—This plant is interesting from the fact that it is parasitic on a mushroom, *Collybia dryophila*. It grows on the stem or on the top of the cap of the *Collybia*, and it is white, or yellowish, very much contorted (gyrose-plicate), nearly rounded, and 8–16 mm. in diameter. Figure 205 represents this *Tremella* growing on the *Collybia dryophila*, from plants collected at Freeville woods near Ithaca.

Figure 206.—Tremella frondosa. Pinkish yellow or pinkish vinaceous (natural size). Copyright.

Tremella frondosa Fr.—This is said to be the largest species of the genus. It grows on rotten wood. It occurs in Europe, has been collected in New York State, and the Fig. 206 is from a plant (No. 4339, C. U. herbarium) collected at Blowing Rock, N. C., in September, 1899. The plant figured here was 10 cm. long and about 8 cm. high. It is very much twisted and contorted, leaf-like, and the middle and base all united. It is of a [Pg 206] pinkish yellow color, one plant being vinaceous pink and another cream buff in color. When young the leaf-like lobes do not show well, but as it expands they become very prominent.

Several other species of Tremella are probably more common than the ones illustrated here. One of the commonest of the *Tremellineæ* probably is the **Exidia glandulosa**, which in dry weather appears as a black incrustation on dead limbs, but during rains it swells up into a large, black, very soft, gelatinous mass. It is commonly found on fallen limbs of oak, and occurs from autumn until late spring. It is sometimes called "witch's butter."

Figure 207. — Tremella fuciformis. Entirely white (natural size). Copyright.

Tremella fuciformis Berk. — This is a very beautiful white tremella growing in woods on leaf mold close to the ground. It forms a large white tubercular mass resting on the ground, from the upper surface of which numerous stout, short, white processes arise which branch a few times in a dichotomous manner. The masses are 10–15 cm. in diameter, and nearly or quite as high. The flesh is very soft, and the parts are more or less hollow. The basidia are like those of the genus, globose, sunk in the substance of the plant, and terminate with four long, slender, sterigmata which rise to the surface and bear the spores. The spores are white, nearly ovoid, but inequilateral and somewhat reniform, continuous, 7–9 × 5–6 μ. [Pg 207]

Figure 207 is from a plant collected in a woods near Ithaca, in August, 1897.

GYROCEPHALUS Pers.

The genus *Gyrocephalus* differs from the other *Tremellineæ* in having the fruiting surface on the lower side of the fruit body, while the upper side is sterile.

Figure 208.—Gyrocephalus rufus. Reddish or reddish yellow (natural size). Copyright.

Gyrocephalus rufus (Jacq.) Bref.—This species is sometimes very abundant. It grows on the ground, generally from buried wood, or from dead roots. It is erect, stout at the base, and the upper end flattened and thinner. It is more or less spatulate, the upper side somewhat concave, and the lower somewhat convex. In some plants the pileus is more regular and there is then a tendency to the funnel form. It is reddish, or reddish yellow in color, smooth, clammy, watery, and quite gelatinous. When dry it is very hard. Figure [Pg 208] 208 represents the form of the plant well, from plants collected at Ithaca. The plant is quite common in the damp glens and woods at Ithaca during the autumn.

CHAPTER XIII.

THELEPHORACEAE.

Many of the species of the Thelephoraceæ to which the following two species belong are too tough for food. A large number of these grow on wood. They are known by their hard or membranaceous character and by the fruiting surface (under surface when in the position in which they grew) being smooth, or only slightly uneven, or cracked.

Craterellus cantharellus (Schw.) Fr., is an edible species. In general appearance it resembles the *Cantharellus cibarius*. The color is the same, and the general shape, except that the former is perhaps more irregular in form. It may, however, be in most cases easily distinguished from *C. cibarius* by the absence of folds on the under or fruiting surface, since the fruiting surface is smooth, especially when the plants are young or middle age. However, when the plants get quite large and old, in some cases the fruiting surface becomes very uneven from numerous folds and wrinkles, which, however, are more irregular than the folds of *C. cibarius*.

Craterellus cornucopioides (L.) Pers., is another edible species. It grows on the ground in woods. It is of a dusky or dark smoky color, and is deeply funnel-shaped, resembling a "horn of plenty," though usually straight. The fruiting surface is somewhat uneven.

The genus *Stereum* is a very common one on branches, etc., either entirely spread out on the wood, or with the margin or a large part of the pileus free. *Hymenochæte* is like *Stereum*, but has numerous small black spines in the fruiting surface, giving it a velvety appearance. *Corticium* is very thin and spread over the wood in patches.

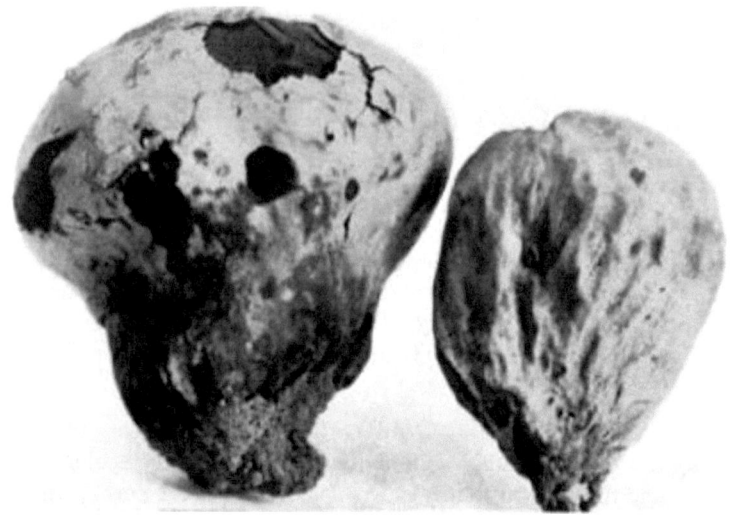

Plate 81, Figure 209.—Lycoperdon cyathiforme (natural size).

CHAPTER XIV. [Pg 209]

PUFF-BALLS: LYCOPERDACEAE.

This is not the place for a discussion of the different genera of the puff-balls, etc., but it might be well to say that in recent years the old genus *Lycoperdon* has been divided into several genera. The giant puff-ball, and the *L. cyathiforme*, where the wall or peridium ruptures irregularly, have been placed in a genus called *Calvatia*; certain other species which are nearly globose, and in which the wall is of a papery texture at maturity, are placed in the genus *Bovista*. There is one genus belonging to the same family as the lycoperdons, the species of which are very interesting on account of the peculiar way in which the wall is ruptured. This is the genus *Geaster*, that is, "earth star." The wall, or peridium, is quite thick in the members of this genus, and when it matures it separates into several layers which need not all be discussed here. A thick outer portion which separates from a thinner inner portion further splits radially into several star-like divisions, which spread outward and give to the plant the form of a star. Since the plants lie on the earth the name earth star was applied to them. This opens out in dry weather, even curving around under the plant, so that the plant is raised above the ground. Then in wet weather it closes up again. The inner portion of the wall opens at the apex in various ways, in the different species, so that the spores may escape. A closely related genus has several small perforations like a pepper box in the upper surface of the inner wall, *Myriostoma*.

LYCOPERDON Tourn.

To this genus belong most of the "puff-balls," as they are commonly called, or, as they are denominated in the South, "Devil's snuff box." All, or a large portion, of the interior of the plant at maturity breaks down into a powdery substance, which with the numerous spores is very light, and when the plant is squeezed or pressed, clouds of this dust burst out at the opening through the wall. The wall of the plant is termed the *peridium*. In this genus the wall is quite thin, and at maturity opens differently in different species. In several species it opens irregularly, the entire wall becoming very brittle and cracking up into bits, as in the giant puff-ball. In the [Pg 210] remaining species it opens by a distinct perforation at the apex, and the remainder of the wall is more or less pliant and membranous. All of the puff-balls are said to be edible, at least are harmless, if eaten when the flesh is white. They should not be eaten when the flesh is dark, or is changing from the white color.

Lycoperdon giganteum Batsch. **Edible.** — This, the giant puff-ball, is the largest species of the genus. Sometimes it reaches immense proportions, two to three or even four feet, but these large sizes are rare. It is usually 20 to 40 cm. (8–16 in.) in diameter. It grows on the ground in grassy places during late summer and in the autumn. It is a large rounded mass, resting on the ground, and near or at the center of the under side, it is attached to the cords of mycelium in the ground. It is white in color until it is ripe, that is, when the spores are mature, and it should be gathered for food before it is thus ripe. When it is maturing it becomes yellowish, then dusky or smoky in color. The flesh, which is white when young, changes to greenish yellow and finally brownish, with usually an olivaceous tinge, as the spores ripen.

The plant is so large that it may be sliced, and should be sliced before broiling. A single specimen often forms enough for a meal for a large family, and some of the larger ones would serve for several meals.

Lycoperdon cyathiforme Bosc. **Edible.**—This is called the beaker-shaped puff-ball because the base of the plant, after the spores have all been scattered, resembles to some extent a beaker, or a broad cup with a stout, stem-like base. These old sterile bases of the plant are often found in the fields long after the spores have disappeared. The plants are somewhat pear-shaped, rounded above, and tapering below to the stout base. They are 7–15 cm. in diameter, and white when young. At maturity the spore mass is purplish, and by this color as well as by the sterile base the plant is easily recognized. Of course these characters cannot be recognized in the young and growing plant at the time it is wanted for food, but the white color of the interior of the plant would be a sufficient guarantee that it was edible, granted of course that it was a member of the puff-ball family. Sometimes, long before the spores mature, the outer portion of the plant changes from white to pinkish, or brownish colors. At maturity the wall, or peridium, breaks into brittle fragments, which disappear and the purplish mass of the spores is exposed. The plant grows in grassy places or even in cultivated fields.

Lycoperdon gemmatum Batsch. **Edible.**—This puff-ball is widely distributed throughout the world and is very common. It grows in the [Pg 211] woods, or in open places on the ground, usually. It is known from its characteristic top shape, the more or less erect scales on the upper surface intermingled with smaller ones, the larger ones falling away and leaving circular scars over the surface, which gives it a reticulate appearance. The plants are white, becoming dark gray or grayish brown when mature. They vary in size from 3–7 cm. high to 2–5 cm. broad. They are more or less top-shaped, and the stem, which is stout, is sometimes longer than the rounded portion, which is the fruiting part. The outer part of the wall (outer peridium) when quite young separates into warts or scales of varying size, large ones arranged quite regularly with smaller ones between. These warts are well shown in the two plants at the left in Fig. 210, and the third plant from the left shows the reticulations formed of numerous scars on the inner peridium where the larger scales have fallen away.

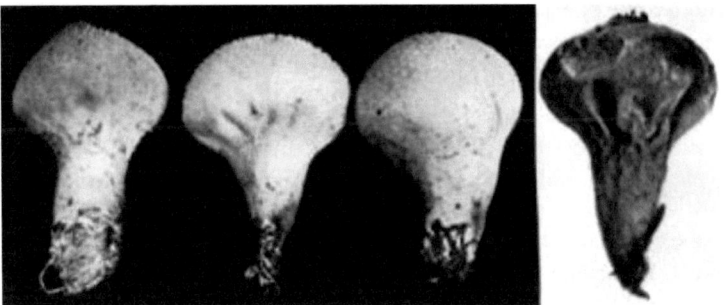

Figure 210.—Lycoperdon gemmatum. Entirely white except when old (natural size). Copyright.

The plant at the extreme right is mature, and the inner peridium has ruptured at the apex to permit the escape of the spores. The spore mass, together with brownish threads which are intermingled, are greenish yellow with an olive tinge, then they become pale brown. The spores are rounded, 3.5–4.5 μ in diameter, smooth or minutely warted.

Another small puff-ball everywhere common in woods is the *Lycoperdon pyriforme*, so called because of its pear shape. It grows on very rotten wood or on decaying logs in woods or groves, or in open places where there is rotting wood. It is somewhat smaller than the gem-bearing lycoperdon, is almost sessile, sometimes many crowded very close together, and especially is it characterized by prominent root-like white strands of mycelium which are attached to the base [Pg 212] where the plant enters the rotten wood. While these small species of puff-balls are not injurious to eat, they do not seem to possess an agreeable flavor. There are quite a number of species in this country which cannot be enumerated here.

Related to the puff-balls, and properly classed with them, are the species of *Scleroderma*. This name is given to the genus because of the hard peridium, the wall being much firmer and harder than in *Lycoperdon*. There are two species which are not uncommon, *Scleroderma vulgare* and *S. verrucosum*. They grow on the ground or on very rotten wood, and are sessile, often showing the root-like white strands attached to their base. They vary in size from 2–6 cm. and the outer wall is cracked into numerous coarse areas, or warts, giving the plant a verrucose appearance, from which one of the species gets its specific name.

Calostoma cinnabarinum Desv. — This is a remarkably beautiful plant with a general distribution in the Eastern United States. It has often been referred to in this country under the genus name *Mitremyces*, and sometimes has been confused with a rarer and different species, *Calostoma lutescens* (Schw.) Burnap. It grows in damp woods, usually along the banks of streams and along mountain roads. It is remarkable for the brilliant vermilion color of the inner surface of the outer layer of the wall (*exoperidium*), which is exposed by splitting into radial strips that curl and twist themselves off, and by the vermilion color of the edges of the teeth at the apex of the inner wall (*endoperidium*). The plant is 2–8 cm. high, and 1–2 cm. in diameter. When mature the base or stem, which is formed of reticulated and anastomosing cords, elongates and lifts the rounded or oval fruiting portion to some distance above the surface of the ground, when the gelatinous volva ruptures and falls to the ground or partly clings to the stem, exposing the peridium, the outer portion of which then splits in the manner described.

When the plant is first seen above the ground it appears as a globose or rounded body, and in wet weather has a very thick gelatinous layer surrounding it. This is the volva and is formed by the gelatinization of the outer layer of threads which compose it. This gelatinous layer is thick and also viscid, and when the plants are placed on paper to dry, it glues them firmly to the sheet. When the outer layer of the peridium splits, it does so by splitting from the base toward the apex, or from the apex toward the base. Of the large number of specimens which I have seen at Blowing Rock, N. C., the split more often begins at the apex, or at least, when the slit is complete, the strips usually stand out loosely in a radiate manner, the [Pg 213] tips being free. At this stage the plant is a very beautiful object with the crown of vermilion strips radiating outward from the base of the fruit body at the top of the stem, and the inner peridium resting in the center and terminated by the four to seven teeth with vermilion edges. At this time also the light yellow spore mass is oozing out from between the teeth. The spores are oblong to elliptical, marked with very fine points, and measure 15–18 × 8–10 μ.

Plate 82, Figure 211.—Calostoma cinnabarinum. See text for colors (natural size).

Figure 211 is from plants collected at Blowing Rock, N. C., in September, 1899. The *Mytremyces lutescens* reported in my list of "Some Fungi of Blowing Rock, N. C.," in Jour. Elisha Mitchell Sci. Soc. 9: 95–107, 1892, is this *Calostoma cinnabarinum*.

CHAPTER XV.

THE STINK-HORN FUNGI: PHALLOIDEAE Fries.

Most of the stink-horn fungi are characterized by a very offensive odor. Some of them at maturity are in shape not unlike that of a horn, and the vulgar name is applied because of this form and the odor. The plants grow in the ground, or in decaying organic matter lying on the ground. The spawn or mycelium is in the form of rope-like strands which are usually much branched and matted together. From these cords the fruit form arises. During its period of growth and up to the maturity of the spores, the fruit body is oval, that is, egg form, and because of this form and the quite large size of these bodies they are often called "eggs." The outer portion of the egg forms the volva. It is always thick, and has an outer thin coat or membrane, and an inner membrane, while between the two is a thick layer of gelatinous substance, so that the wall of the volva is often 3–6 mm. in thickness, and is very soft. The outline of the volva can be seen in Fig. 215, which shows sections of three eggs in different stages. Inside of the volva is the short stem (*receptacle*) which is in the middle portion, and covering the upper portion and sides of this short stem is the pileus; the fruit-bearing portion, which is divided into small chambers, lies on the outside of the pileus. In the figure there can be seen cross lines extending through this part from the pileus to the wall of the volva. These represent ridges or crests which anastomose over the pileus, forming reticulations. The stem or receptacle is hollow through the center, and this hollow opens [Pg 214] out at the end so that there is a rounded perforation through the upper portion of the pileus.

The spores are borne on club-shaped basidia within the chambers of the fruit-bearing portion (*gleba*), and at maturity of the spores the stem or receptacle begins to elongate. This pushes the gleba and the upper part of the receptacle through the apex of the volva, leaving this as a cup-shaped body at the base, much as in certain species of *Amanita*, while the gleba is borne aloft on the much elongated stem. During this elongation of the receptacle a large part of the substance of the gleba dissolves into a thick liquid containing the spores. This runs off and is washed off by the rains, leaving the inner surface of the gleba exposed, and showing certain characters peculiar to the various genera.

Among the stink-horns are a number of genera which are very interesting from the peculiarities of development; and some of which are very beautiful and curious objects, although they do possess offensive odors. In some of the genera, the upper part of the plant expands into leaf-like — or petal-like, bodies, which are highly colored and resemble flowers. They are sometimes called "fungus flowers."

DICTYOPHORA Desvaux.

Dictyophora means "net bearer," and as one can see from Fig. 212 it is not an inappropriate name. The stem or receptacle, as one can see from the illustrations of the two species treated of here, possesses a very coarse mesh, so that not only the surface but the substance within is reticulated, pitted and irregularly perforated. In the genus *Dictyophora* an outer layer of the receptacle or stem is separated as it elongates, breaks away from the lower part of the stem, is carried aloft, and hangs as a beautiful veil. This veil is very conspicuous in some species and less so in others.

Dictyophora duplicata (Bosc.) Ed. Fischer.—This species is illustrated in Fig. 212, made from plants collected at Ithaca. The plants are from 15–22 cm. high, the cap about 5 cm. in diameter, and the stem 2–3 cm. in thickness. According to Burt (Bot. Gaz. **22**: 387, 1896) it is a common species in the Eastern United States. The cap is more or less bell-shaped and the sculptured surface is marked in a beautiful manner with the reticulations.

Plate 83, Figure 212.—Dictyophora duplicata. White (natural size). Copyright.

[Pg 215]

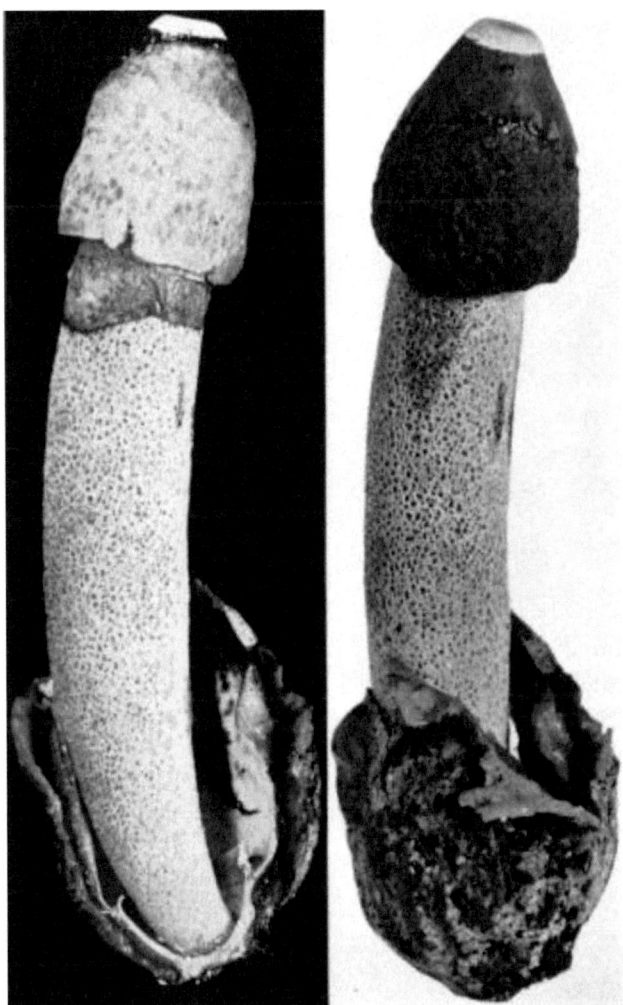

Plate 84, Figure 213.—Dictyophora ravenelii. Mature plants showing volva at base; elongated receptacle, cap at the top, and veil surrounding the receptacle under the cap (natural size). Copyright.

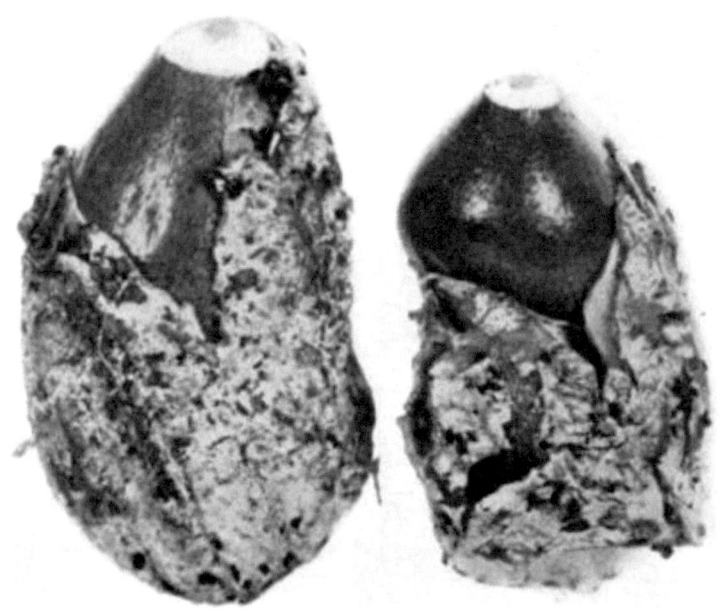

Figure 214. — Dictyophora ravenelii. Egg stage, caps just bursting through the volva (natural size). Copyright.

Figure 215. — Dictyophora ravenelii. Sections of eggs, and showing cords of mycelium (natural size). Copyright.

Dictyophora ravenelii (B. & C.) Burt.—This plant also has a wide distribution in the Eastern United States. The stem is more slender than in the other species, *D. duplicata*, the pileus more nearly conic, and the surface of the pileus is merely granular or minutely wrinkled after the disappearance of the gleba, and does not present the strong reticulating ridges and crests which that species shows. The plants are from 10 to 18 cm. high. It grows in woods and fields about rotting wood, and in sawdust. The veil is very thin and delicate, forming [Pg 216] simply a membrane, and does not possess the coarse meshes present in the veil of *D. duplicata*. The Figs. 214, 215 represent the different stages in the elongation of the receptacle of this plant, and the rupture of the volva. This elongation takes place quite rapidly. While photographing the plant as it was bursting through the volva, I had considerable difficulty in getting a picture, since the stem elongated so rapidly that the plant would show that it had moved perceptibly, and the picture would be blurred.

In a woods near Ithaca a large number of these plants have appeared from year to year in a pile of sawdust. One of the most vile smelling plants of this family is the *Ithyphallus impudicus*.

CHAPTER XVI.

MORELS, CUP-FUNGI, HELVELLAS, ETC.: DISCOMY-CETES.

The remaining fungi to be considered belong to a very different group of plants than do the mushrooms, puff-balls, etc. Neverthel-ess, because of the size of several of the species and the fact that several of them are excellent for food, some attention will be given to a few. The entire group is sometimes spoken of as *Discomycetes* or *cup-fungi*, because many of the plants belonging here are shaped something like a disk, or like a cup. The principal way in which they differ from the mushrooms, the puff-balls, etc., is found in the manner in which the spores are borne. In the mushrooms, etc., the spores, we recollect, are borne on the end of a club-shaped body, usually four spores on one of these. In this group, however, the spores are borne inside of club-shaped bodies, called sacs or asci (singular, ascus). These sacs, or asci, are grouped together, lying side by side, forming the fruiting surface or hymenium, much as the basidia form the fruiting surface in the mushrooms. In the case of the cup or disk forms, the upper side of the disk, or the upper and inner surface of the cap, is covered with these sacs, standing side by side, so that the free ends of the sacs form the outer surface. In the case of the morel the entire outer surface of the upper portion of the plant, that where there are so many pits, is covered with similar sacs. Since so few of the genera and species of the morels and cup-fungi will be treated of here, I shall not attempt to compare the genera or even to give the characters by which the genera are known. In [Pg 217] most cases the illustrations will serve this purpose so far as it is desirable to accomplish it in such a work as the present. Certain of the species will then be described and illustrated.

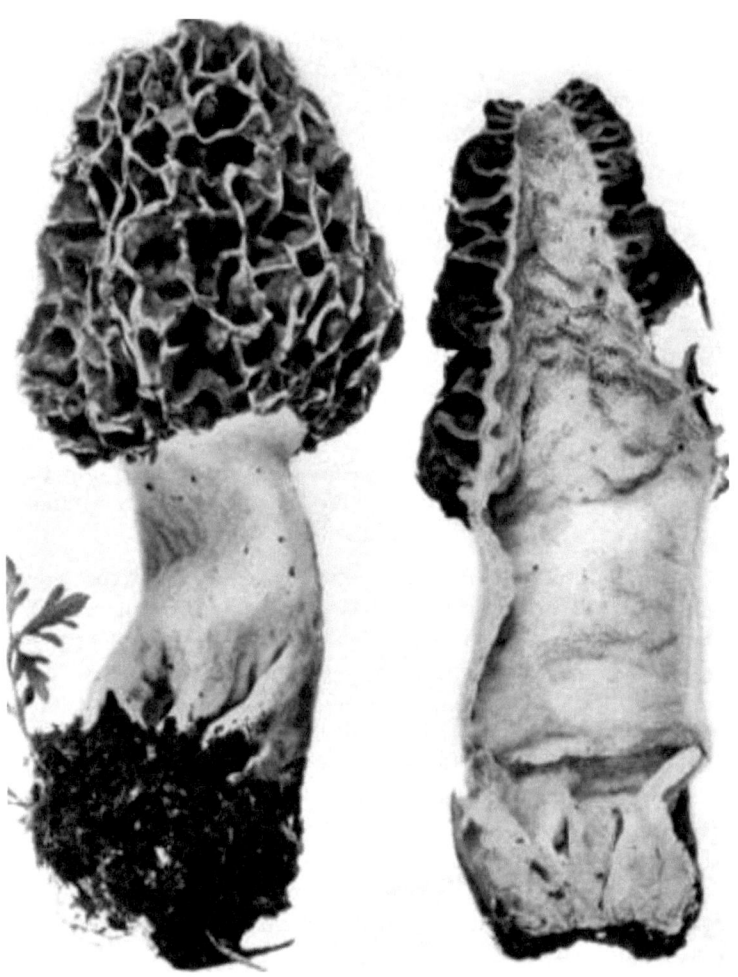

Plate 85, Figure 216.—Morchella esculenta (natural size). Copyright.

MORCHELLA Dill.

The morels are all edible and they are usually easy to recognize. The plant consists of two distinct, prominent parts, the cap and the stem. The cap varies in form from rounded, ovate, conic or cylindrical, or bell-shaped, but it is always marked by rather broad pits, covering the entire outer surface, which are separated from each other by ridges forming a network. The color of the plants does not differ to any perceptible extent in our species. The cap is usually buff or light ochre yellow, becoming darker with age and in drying.

The stem in all our species is usually quite stout, though it varies to some extent in some of the different species, in proportion to the thickness of the cap. The stem is marked in some of the species by large wrinkles or folds extending irregularly but with considerable uniformity over the surface. The surface is further minutely roughened by whitish or grayish elevations, giving it a granular appearance. Sometimes these granules are quite evenly distributed over the surface, and in some species they are more or less separated into small areas by narrow lines.

The morels appear early in the season, during May and June. They grow usually in damp situations, and are more abundant during rainy weather. Three species are illustrated here.

Morchella esculenta Pers. **Edible.** — The name of this species, the esculent morel, indicates that it has been long known as an edible plant, especially since the man who named it lived a century ago. The plant is from 5–15 cm. high, the stem is 1–3 cm. in thickness, and the cap is broader than the stem. The cap is somewhat longer than broad, and is more or less oval or rounded in outline. The arrangement of the pits on the surface of the cap is regarded by some as being characteristic of certain species. In this species the pits are irregularly arranged, so that they do not form rows, and so that the ridges separating them do not run longitudinally from the base toward the apex of the cap, but run quite irregularly. This arrangement can be seen in Fig. 216, which is from a photograph of this species. The stem is hollow.

Morchella conica Pers. **Edible.** — This species is very closely related to the preceding one, and is considered by some to be only a form of the *Morchella esculenta*. The size is about the same, the only difference being in the somewhat longer cap and especially in the arrangement [Pg 218] of the pits. These are arranged more or less in distinct rows, so that the ridges separating them run longitudinally and parallel from the base of the cap to the apex, with connecting ridges extending across between the pits. The cap is also more or less conic, but not necessarily so. Figure 217 illustrates this species. The plant shown here is branched, and this should not be taken to be a character of the species, for it is not, this form being rather rare.

Figure 217. — Morchella conica (natural size). Copyright.

[Pg 219]

Plate 86, Figure 218.—Morchella crassipes (natural size). Copyright.

Morchella crassipes (Vent.) Pers. **Edible.**—This species differs from the two preceding in the fact that the stem is nearly equal in width with the cap. Figure 218 illustrates a handsome specimen which was 17 cm. high. The granular surface and the folds of the stem show very distinctly and beautifully. Collected at Ithaca.

Morchella deliciosa Fr. **Edible,** has the cap cylindrical or nearly so. It is longer than the stem, and is usually two or three times as long as it is broad. The plant is smaller than the preceding, though large ones may equal in size small ones of those two. The plant is from 4–8 cm. high.

Morchella semilibera DC., and **M. bispora** Sor., [*Verpa bohemica* (Kromb.) Schroet.] occur in this country, and are interesting from the fact that the cap is bell-shaped, the lower margin being free from the stem. In the latter species there are only two spores in an ascus.

HELVELLA L.

The helvellas are pretty and attractive plants. They are smaller than the morels, usually. They have a cap and stem, the cap being very irregular in shape, often somewhat lobed or saddle-shaped. It is smooth, or nearly so, at least it is not marked by the large pits present in the cap of the morel, and this is one of the principal distinguishing features of the helvellas as compared with the morels. In one species the thin cap has its lower margin free from the stem. This is **Helvella crispa** Fr., and it has a white or whitish cap, and a deeply furrowed stem. It occurs in woods during the summer and autumn, and is known as the white helvella.

[Pg 220]

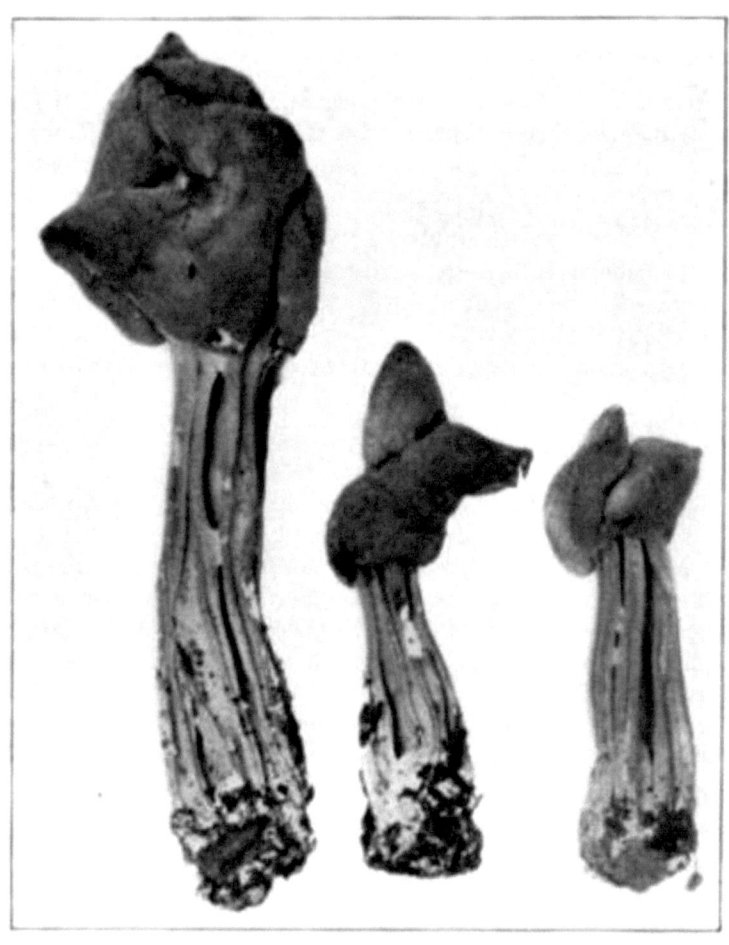

Figure 219. — Helvella lacunosa (natural size). Copyright.

Another species which has a wide range is the **Helvella lacunosa**, so called because of the deep longitudinal grooves in the stem. The cap is thin, but differs from the *H. crispa* in that the lower margin is connected with the stem. This species is illustrated in Fig. 219 from plants collected at Blowing Rock, N. C., during September, 1899.

The genus *Gyromitra* is very closely related to *Helvella*, and is only distinguished by the fact that the cap is marked by prominent folds and convolutions, resembling somewhat the convolutions of the brain. Its name means *convoluted cap*. The <u>Gyromitra esculenta</u> Fr., is from 5–10 cm. high, and the cap from 5–7 cm. broad. While this species has long been reported as an edible one, and has been employed in many instances as food with no evil results, there are known cases where it has acted as a poison. In many cases where poisoning has resulted the plants were quite old and probably in the incipient stages of decay. However, it is claimed that a poisonous principle, called *helvellic acid*, has been isolated by a certain chemist, which acts as a violent poison. This principle is very soluble in hot water, and when care is used to drain off first water in which they have been cooked, squeezing the water well from the plants, they are pronounced harmless. The safer way would be to avoid such suspicious species.

Figure 220.—Spathularia velutipes (natural size). Copyright.

Spathularia velutipes Cooke & Farlow. — This species represents another interesting genus of the *Discomycetes*. It is in the form of a "spatula," and from this shape of the plant the genus takes its name. There are several species known in this country, and this one is quite [Pg 221] common. The stem extends the entire length of the plant, running right through the cap, or perhaps it would be better to say that the cap or fruiting portion forms two narrow blades or wings on opposite sides of the upper part of the stem. These wing-like expansions of the cap on the opposite sides of the stem give the spathulate form to the plant. Figure 220 is from plants collected in the woods near Ithaca.

Figure 221. — Leotia lubrica (natural size). Copyright.

Leotia lubrica Pers. — The genus *Leotia* is quite readily recognized by its form, and because the plants are usually slimy. This species is called *lubrica* because of the slippery character of the entire plant. It is dull yellowish or olive yellow in color. The cap, as can be seen from the figure (221), is irregularly rounded, and broader than the stem. The plant is illustrated natural size from specimens collected near Ithaca.

THE TRUE CUP-FUNGI.

By far the larger number of the *Discomycetes* are cup-shaped, and are popularly called "cup-fungi." They vary from plants of very minute size, so small that they can be just seen with the eye, or some of the larger ones are several inches in breadth. They grow on the ground, on leaves, wood, etc. The variety of form and color is great.

They may be sessile, that is, the cup rests immediately on the ground or wood, or leaves, or they may possess a short, or rather long stalk. The only species illustrated here has a comparatively long stalk, and the cap is deep cup-shaped, almost like a beaker. This plant is technically known as *Sarcoscypha floccosa*. It is represented [Pg 222] here natural size (Fig. 222). The stem is slender, and the rim of the cup is beset with long, strigose hairs. The inner surface of the cup is lined with the sacs (asci) and sterile threads (paraphyses), spoken of on a former page, when treating of the fruiting character of the morels and cup-fungi. In this plant the color of the inside of the cup is very beautiful, being a bright red. Another species, *Sarcoscypha coccinea*, the scarlet sarcoscypha, is a larger plant which appears in very early spring, soon after the frost is out of the ground. It grows on rotting logs and wood in the woods or in groves. The inside of the cup in this species is a rich scarlet, and from this rich color the species takes its name.

Figure 222.—Sarcoscypha floccosa (natural size). Copyright.

CHAPTER XVII.

COLLECTION AND PRESERVATION OF THE FLESHY FUNGI.

In the collection of the higher fungi it is of the utmost importance that certain precautions be employed in obtaining all parts of the plant, and furthermore that care be exercised in handling, in order not to remove or efface delicate characters. Not only is it important for the beginner, but in many instances an "expert" may not be able to determine a specimen which may have lost what undoubtedly seem, to some, trivial marks. The suggestions given here should enable one to collect specimens in such a way as to protect these characters while fresh, to make notes of the important evanescent characters and to dry and preserve them properly for future study. For collecting a number of specimens under a variety of conditions the following list of "apparatus" is recommended: [Pg 223]

One or two oblong or rectangular hand baskets, capacity from 8–12 quarts.

Or a rectangular zinc case with a closely fitting top (not the ordinary botanical collecting case).

Half a dozen or so tall pasteboard boxes, or tins, 3 × 3, or 4 × 4, × 5 inches deep, to hold certain species in an upright position.

A quantity of tissue paper cut 8 × 10 or 6 × 8 inches.

Smaller quantity of waxed tissue paper for wrapping viscid or sticky plants.

Trowel; a stout knife; memorandum pad and pencil.

Collecting.—During the proper season, and when rains are a-bundant, the mushrooms are to be found in open fields, waste places, groves and woods. They are usually more abundant in the forests. Especially in dry weather are specimens more numerous in rather damp woods, along ravines or streams. In collecting specimens which grow on the ground the trowel should be used to dig up the plant carefully, to be sure that no important part of the plant is left in the ground. After one has become familiar with the habit of the different kinds the trowel will not be necessary in all cases. For example, most species of *Russula*, *Lactarius*, *Tricholoma*, *Boletus*, etc., are not deeply seated in the soil, and careful hand-picking will in most cases secure specimens properly, especially if one does not object to digging in the soil with the fingers. But in the case of most species of *Amanita*, certain species of *Lepiota*, *Collybia*, etc., a trowel is necessary to get up the base of the plant in such a way as to preserve essential characters. Even then it is possible, if the ground is not too hard, to dig them out with the fingers, or with a stout knife, but I have often found specimens which could only be taken up with a trowel or spade.

Species growing on sticks or leaves are easily collected by taking a portion of the substratum on which they grow. Specimens on the larger limbs or trunks or stumps can sometimes be "picked," but until one is accustomed to certain individualities of the plant it is well to employ the knife and to cut off a portion of the wood if necessary, to avoid cutting off the base of the stem.

It is necessary also to handle the specimens with the greatest care to avoid leaving finger marks where the surface of the stem or cap is covered with a soft and delicate outer coat, especially if one wishes to photograph the plant, since rubbed or marked places spoil the plant for this purpose. Also a little careless handling will remove such important characters as a frail annulus or volva, which often are absolutely necessary to recognize the species. [Pg 224]

Having collected the specimens, they should be properly placed in the basket or collecting case. Those which are quite firm, and not long and slender, can be wrapped with tissue paper (waxed tissue paper if they are viscid or sticky), and placed directly in the basket, with some note or number to indicate habitat or other peculiarity which it is desirable to make at the time of collection. The smaller, more slender and fragile, specimens can be wrapped in tissue paper (a cluster of several individuals can be frequently rolled up together) made in the form of a narrow funnel and the ends then twisted. The shape of the paper enables one to wrap them in such a way as to protect certain delicate characters on the stem or cap. These can then be stood upright in the small pasteboard boxes which should occupy a portion of the basket. A number of such wrappers can be placed in a single box, unless the specimens are of considerable size and numerous. In these boxes they are prevented from being crushed by the jostling of the larger specimens in the basket. These boxes have the additional advantage of preserving certain specimens entire and upright if one wishes later to photograph them.

Field Notes.—The field notes which may be taken upon the collection will depend on circumstances. If one goes to the sorting room soon after the collection is made, so that notes can be made there before the more delicate specimens dry, few notes will answer in the field, and usually one is so busy collecting or hunting for specimens there is not much inclination to make extended notes in the field. But it is quite important to note the *habitat* and *environment*, i. e., the place where they grow, the kind and character of the soil, in open field, roadside, grove, woods, on ground, leaves, sticks, stumps, trunks, rotting wood, or on living tree, etc. It is very important also that different kinds be kept separate. The student will recognize the importance of this and other suggestions much more than the new "fungus hunter."

Sorting Room. — When one returns from a collecting trip it is best to take the plants as soon as possible to a room where they can be assorted. An hour or so delay usually does not matter, but the sooner they are attended to the better. Sometimes when they are carefully placed in the basket, as described above, they may be kept over night without injury, but this will depend on the *kinds* in the collection. *Coprini* are apt to deliquesce, certain other specimens, especially in warm weather, are apt to be so infested with larvæ that they will be ruined by morning, when immediate drying might save them. Other thin and delicate ones, especially in dry weather, will dry out so completely that one loses the opportunity of taking [Pg 225] notes on the fresh specimen. Specimens to be photographed should be attended to at once, unless it is too late in the day, when they should be set aside in an upright position, and if necessary under a bell-jar, until the following day. As far as possible good specimens should be selected for the photograph, representing different stages of development, and one to show the fruiting surface. Sometimes it will be necessary to make more than one photograph to obtain all the stages. Also on different days one is apt to obtain a specimen representing an important stage in development not represented before. The plants should be arranged close together to economize space, but not usually touching nor too crowded. They should be placed in their natural position as far as possible, and means for support, if used, should be hidden behind the plant. They should be so arranged as to show individual as well as specific character and should be photographed if possible natural size, or at least not on a plate smaller than 5 × 7 inches unless the plants are small; while larger ones are better on 6 × 8 or larger. Some very small ones it may be necessary to enlarge in order to show the character of the fruiting surface, and even large specimens can sometimes have a portion of the hymenium enlarged to good advantage if it is desirable to show the characters clearly. The background should be selected to bring out the characters strongly, and in the exposure and developing it is often necessary to disregard the effect of the background in order to bring out the detail of texture on the plant itself. The background should be renewed as often as necessary to have it uniform and neat. There is much more that might be said under this head, but there is not space here.

To Obtain Spore Prints. — In many cases it is desirable to obtain spores in a mass on paper in order to know the exact tint of color produced by the species. Often the color of the spores can be satisfactorily determined by an examination of them under the microscope. One cannot always depend on the color of the lamellæ since a number of the species possess colored cystidia or spines in the hymenium which disguise the color of the spores. The best way to determine the color of the spores in mass is to catch them as they fall from the fruiting surface on paper. For the ordinary purpose of study and reference in the herbarium the spores caught on unprepared paper, which later may be placed in the packet with the specimen, will answer. This method has the advantage of saving time, and also the danger of injury to the spores from some of the fixatives on prepared paper is avoided. If for purposes [Pg 226] of illustration one wishes pretty spore prints, perfect caps must be cut from the stem and placed fruiting surface downward on paper prepared with some gum arable or similar preparation spread over it, while the paper is still moist with the fixative, and then the specimen must be covered with a bell-jar or other receiver to prevent even the slightest draft of air, otherwise the spores will float around more or less. The spores may be caught on a thin, absorbent paper, and the paper then be floated on the fixative in a shallow vessel until it soaks through and comes in contact with the spores. I have sometimes used white of egg as a fixative. These pieces of paper can then be cut out and either glued to card-boards, or onto the herbarium sheet.

Sorting the Plants. — This should be done as soon as possible after collection. A large table in the sorting room is convenient, upon which the specimens may be spread, or grouped rather, by species, the individuals of a species together, on sheets of paper. Surplus dirt, or wood, leaves, etc., can be removed. A few of the specimens can be turned so that spores can be caught on the papers. If only one or a few specimens of a given species have been found, and it is desirable not to cut off the cap from the stem, the plant can be supported in an upright position, a small piece of paper slit at one side can be slipped around the stem underneath the cap, on which the spores will fall. Sometimes it will be necessary to cover the plant with a bell-jar in order to prevent it from drying before the spores are shed. Experience with different species will suggest the treatment necessary.

Taking Notes on the Specimens.—Very few probably realize the desirability of making notes of certain characters while the plants are fresh, for future reference, or for use by those to whom the plants may be sent for determination. It is some trouble to do this, and when the different kinds are plentiful the temptation is strong to neglect it. When one has available books for determination of the species, as many as possible should be studied and determined while fresh. But it is not always possible to satisfactorily determine all. Some may be too difficult for ready recognition, others may not be described in the books at hand, or poorly so, and further the number of kinds may be too great for determination before they will spoil. On these as well as on some of the interesting ones recognized, it is important to make a record of certain characters. These notes should be kept either with the specimen, or a number should be given the specimen and the notes kept separately with the corresponding number.

MEMORANDA. [Pg 227]

No.____. Locality, Date. Name of collector.

Weather.

Habitat.—If on ground, low or high, wet or dry, kind of soil; on fallen leaves, twigs, branches, logs, stumps, roots, whether dead or living, kind of tree; in open fields, pastures, etc., woods, groves, etc., mixed woods or evergreen, oak, chestnut, etc.

Plants.—Whether solitary, clustered, tufted, whether rooting or not, taste, odor, color when bruised or cut, and if a change in color takes place after exposure to the air.

Cap.—Whether dry, moist, watery in appearance (hygrophanous), slimy, viscid, glutinous; color when young, when old; whether with fine bloom, powder; kind of scales and arrangement, whether free from the cuticle and easily rubbed off. Shape of cap.

Margin of Cap.—Whether straight or incurved when young, whether striate or not when moist.

Stem.—Whether slimy, viscid, glutinous, kind of scales if not smooth, whether striate, dotted, granular, color; when there are several specimens test one to see if it is easily broken out from the cap, also to see if it is fibrous, or fleshy, or cartilaginous (firm on the outside, partly snapping and partly tough). Shape of the stem.

Gills or Tubes.—Color when young, old, color when bruised, and if color changes, whether soft, waxy, brittle, or tough; sharp or blunt, plane or serrate edge.

Milk.—Color if present, changing after exposure, taste.

Veil.—(Inner veil.) Whether present or not, character, whether arachnoid, and if so whether free from cuticle of pileus or attached only to the edge; whether fragile, persistent, disappearing, slimy, etc., movable, etc.

Ring.—Present or absent, fragile, or persistent, whether movable, viscid, etc.

Volva.—Present or absent, persistent or disappearing, whether it splits at apex or is circumscissile, or all crumbly and granular or floccose, whether the part on the pileus forms warts, and then the kind, distribution, shape, persistence, etc.

Spores.—Color when caught on white paper.

To the close observer additional points of interest will often be noted.

To Dry the Specimens.—Frequently the smaller specimens will dry well when left in the room, especially in dry weather, or better if [Pg 228] they are placed where there is a draft of air. Some dry them in the sun. But often the sun is not shining, and the weather may be rainy or the air very humid, when it is impossible to dry the specimens properly except by artificial heat. The latter method is better for the larger specimens at all times. During the autumn when radiators are heated the fungi dry well when placed on or over them. One of the best places which I have utilized is the brick work around a boiler connected with a mountain hotel. Two other methods are, however, capable of wider application.

1st. — A tin oven about 2 × 2 feet, and two or several feet high, with one side hinged as a door, and with several movable shelves of perforated tin, or of wire netting; a vent at the top, and perforations around the sides at the bottom to admit air. The object being to provide for a constant current of air from below upwards between the specimens. This may be heated, if not too large, with a lamp, though an oil stove or gas jet or heater is better. The specimens are placed on the shelves with the accompanying notes or numbers. The height of this box can be extended where the number of specimens is great.

2d. — A very successful method which I employed at a summer resort at Blowing Rock, N. C., in the mountains of North Carolina, during September, 1899, was as follows: An old cook stove was set up in an unoccupied cottage, with two wire screens from 3 × 4 feet, one above the other, the lower one about one foot above the top of the stove. Large numbers can be dried on these frames. Care of course must be taken that the plants are not burned. In all cases the plants must be so placed that air will circulate under and around them, otherwise they are apt to blacken.

When the plants are dry they are very brittle and must be handled carefully. When removed from the drier many kinds soon absorb enough moisture to become pliant so that they are not easily broken. Others remain brittle. They may be put away in small boxes; or pressed out nearly flat, *not so as to crush the gills*, and then put in paper packets. The plants which do not absorb sufficient moisture from the air, so that they are pliant enough to press, can be placed in small boxes or on paper in a large box with peat moss in the bottom, and the box then closed tightly until they absorb enough moisture to become flexible. The plants must not get wet, and they should be examined every half hour or so, for some become limp much sooner than others. If the plants get too moist the gills crush together when pressed, and otherwise they do not make such good specimens. When the specimens are dried and placed in the herbarium [Pg 229] they must be protected from insects. Some are already infested with insects which the process of drying does not kill. They must be either poisoned with corrosive sublimate in alcohol, or fumigated with carbon disulphide, and if the latter it must be repeated one or two times at an interval of a month to catch those which were in the egg state the first time. When placed in the herbarium or in a box for storage, naphtha balls can be placed with them to keep out insects, but it should be understood that the naphtha balls will not kill or drive away insects already in the specimens. Where there are enough duplicates, some specimens preserved in 75 per cent. alcohol, under the same number, are of value for the study of structural characters.

CHAPTER XVIII.

SELECTION AND PREPARATION OF MUSHROOMS FOR THE TABLE.

In the selection of mushrooms to eat, great caution should be employed by those who are not reasonably familiar with the means of determination of the species, or those who have not an intimate acquaintance with certain forms. Rarely should the beginner be encouraged to eat them upon his own determination. It is best at first to consult some one who knows, or to send first specimens away for determination, though in many cases a careful comparison of the plant with the figures and descriptions given in this book will enable a novice to recognize it. In taking up a species for the first time it would be well to experiment cautiously.

No Certain Rule to Distinguish the Poisonous from the Edible. — There is no certain test, like the "silver spoon test," which will enable one to tell the poisonous mushroom from the edible ones. Nor is the presence of the so-called "death cup" a sure sign that the fungus is poisonous, for the *Amanita cæsarea* has this cup. For the beginner, however, there are certain general rules, which, if carefully followed, will enable him to avoid the poisonous ones, while at the same time necessarily excluding many edible ones.

1st. — Reject all fungi which have begun to decay, or which are infested with larvæ.

2d.—Reject all fungi when in the button stage, since the characters are not yet shown which enable one to distinguish the genera and species. Buttons in pasture lands which are at the surface [Pg 230] of the ground and not deep-seated in the soil, would very likely not belong to any of the very poisonous kinds.

3d.—Reject all fungi which have a cup or sac-like envelope at the base of the stem, or which have a scaly or closely fitting layer at the base of the stem, and rather loose warts on the pileus, especially if the gills are white. *Amanita cæsarea* has a sac-like envelope at the base of the stem, and yellow gills as well as a yellow cap, and is edible. *Amanita rubescens* has remnants of a scaly envelope on the base of the stem and loose warts on the cap, and the flesh where wounded becomes reddish. It is edible. (See plate 19.)

4th.—Reject all fungi with a milky juice unless the juice is reddish. Several species with copious white milk, sweet or mild to the taste, are edible (see *Lactarius volemus* and *corrugis*).

5th.—Reject very brittle fungi with gills nearly all of equal length, where the flesh of the cap is thin, especially those with bright caps.

6th.—Reject all Boleti in which the flesh changes color where bruised or cut, or those in which the tubes have reddish mouths, also those the taste of which is bitter. *Strobilomyces strobilaceus* changes color when cut, and is edible.

7th.—Reject fungi which have a cobwebby veil or ring when young, and those with slimy caps and clay-colored spores.

In addition, proceed cautiously in all cases, and make it a point to become very familiar with a few species first, and gradually extend the range of species, rather than attempt the first season to eat a large number of different kinds.

All puff-balls are edible so long as they are white inside, though some are better than others. All coral-like or club fungi are edible.

To Clean and Prepare the Specimens.—The mushrooms having been collected, all tough stems, the parts to which earth clings, should be removed. After the specimens are selected, if there is danger that some of them may be infested with larvæ, it is well to cut off the stem close to the cap, for if the insects are in the stem and have not yet reached the cap they may thus be cast away. Some recommend that the tubes of all Boleti be removed, since they are apt to make a slimy mass in cooking.

Where the plants are small they may be cooked entire. Large ones should be quartered, or cut, or sliced, according to the size and form of the plant, or method of cooking.

[Pg 231] CHAPTER XIX.

USES OF MUSHROOMS. [C]

The most prominent and at present important use of mushrooms from the standpoint of the utilitarian is as an article of food. We have now learned that their food value as a nutrient substance is not so great as has been fondly supposed, but, as Mr. Clark points out in Chapter XXII, in addition to the value they certainly do possess as food, they have very great value as condiments or food accessories, and "their value as such is beyond the computation of the chemist or physiologist. They are among the most appetizing of table delicacies, and add greatly to the palatability of many foods when cooked with them." Mushrooms undoubtedly possess a food value beyond that attributed to them by the chemist or physiologist, since it is not possible in laboratory analysis to duplicate the conditions which exist in the natural digestion and assimilation of foods.

Probably the larger number of persons, in America, at present interested in mushrooms, are chiefly concerned with them as an article of food, but a great many of these persons love to tramp to the fields and woods in quest of them just as the sportsman loves to hunt his game with dog and gun. It is quite likely that there will always be a large body of persons who will maintain a lively interest in the collection of *game* mushrooms for food. There are several reasons for this. The zest of the search, the pleasure of discovery, and the healthfulness of the outdoor recreation lend an appetizing flavor to the fruits of the chase not to be obtained by purchasing a few pounds of cultivated mushrooms on the market. It cultivates powers of observation, and arouses a sympathetic feeling toward nature, and with those outdoor environments of man which lend themselves so happily in bettering and brightening life, as well as in prolonging it.

Many others are discovering that the observation of form and habits of mushrooms is a very interesting occupation for those who have short periods of time at their disposal weekly. It requires but a little observation to convince one that there is an interesting variety of form among these plants, that their growth and expansion [Pg 232] operate in conformity with certain laws which result in great variation in form and habit of the numerous kinds on the ground, on leaves, on branches, on tree trunks, etc.

Another very favorable indication accompanying the increasing interest in the study of these plants, is the recognition of their importance as objects for nature study. There are many useful as well as interesting lessons taught by mushrooms to those who stop to read their stories. The long growth period of the spawn in the ground, or in the tree trunk, where it may sometimes be imprisoned for years, sometimes a century, or more, before the mushroom appears, is calculated to dispel the popular notion that the mushroom "grows in a night." Then from the button stage to the ripe fruit, several days, a week, a month, or a year may be needed, according to the kind, while some fruiting forms are known to live from several to eighty or more years. The adjustment of the fruit cap to a position most suitable for the scattering of the spores, the different ways in which the fruit cap opens and expands, the different forms of the fruit surface, their colors and other peculiarities, suggest topics for instructive study and observation. The inclination, just now becoming apparent, to extend nature study topics to include mushrooms is an evidence of a broader and more sympathetic attitude toward nature.

A little extension of one's observation on the habits of these plants in the woods will reveal the fact that certain ones are serious enemies of timber trees and timber. It is quite easy in many cases for one possessing no technical knowledge of the subject to read the story of these "wood destroying" fungi in the living tree. Branches broken by snow, by wind, or by falling timber provide entrance areas where the spores, lodging on the heart wood of broken timber, or on a bruise on the side of the trunk which has broken through the living part of the tree lying just beneath the bark, provide a point for entrance. The living substance (*protoplasm*) in the spawn exudes a "juice" (*enzyme*) which dissolves an opening in the wood cells and permits the spawn to enter the heart of the tree, where decay rapidly proceeds as a result. But very few of these plants can enter the tree when the living part underneath the bark is unbroken.

These observations suggest useful topics for thought. They suggest practical methods of prevention, careful forestry treatment and careful lumbering to protect the young growth when timber trees are felled. They suggest careful pruning of fruit and shade trees, [Pg 233] by cutting limbs smooth and close to the trunk, and then painting the smooth surface with some lead paint.

While we are thus apt to regard many of the mushrooms as enemies of the forest, they are, at the same time, of incalculable use to the forest. The mushrooms are nature's most active agents in the disposal of the forest's waste material. Forests that have developed without the guidance of man have been absolutely dependent upon them for their continued existence. Where the species of mushrooms are comparatively few which attack living trees, there are hundreds of kinds ready to strike into fallen timber. There is a degree of moisture present on the forest floor exactly suited to the rapid growth of the mycelium of numbers of species in the bark, sap wood, and heart wood of the fallen trees or shrubs. In a few years the branches begin to crumble because of the disorganizing effect of the mycelium in the wood. Other species adapted to growing in rotting wood follow and bring about, in a few years, the complete disintegration of the wood. It gradually passes into the soil of the forest floor, and is made available food for the living trees. How often one notices that seedling trees and shrubs start more abundantly on rotting logs.

The fallen leaves, too, are seized upon by the mycelium of a great variety of mushrooms. It is through the action of the mycelium of mushrooms of every kind that the fallen forest leaves, as well as the trunks and branches, are converted into food for the living trees. The fungi, are, therefore, one of the most important agents in providing available food for the virgin forest.

The spawn of some fungi in the forest goes so far, in a number of cases, as to completely envelop those portions of the roots of certain trees as to prevent the possibility of the roots taking up food material and moisture on their own account. In such cases, the oaks, beeches, hornbeams, and the like, have the younger parts of their roots completely enveloped with a dense coat of mycelium. The mycelium in these cases absorbs the moisture from the soil or forest floor and conveys it over to the roots of the tree, and in this way supplies them with both food and water from the decaying humus, the oak being thus dependent on the mycelium. In the fields, however, where there is not the abundance of humus and decaying leaves present in the forest, the coating of mycelium on the roots of these trees is absent, and in this latter case the young roots are provided with root hairs which take up the moisture and food substances from the soil in the ordinary way.

The mushrooms also prevent the forest from becoming choked [Pg 234] or strangled by its own fallen members. Were it not for the action of the mushroom mycelium in causing the decay of fallen timber in the forest, in time it would be piled so high as to allow only a miserable existence to a few choked individuals. The action of the mushrooms in thus disposing of the fallen timber in the forests, and in converting dead trees and fallen leaves into available food for the living ones, is probably the most important role in the existence of these plants. Mushrooms, then, are to be given very high rank among the natural agencies which have contributed to the good of the world. When we contemplate the vast areas of forest in the world we can gain some idea of the stupendous work performed by the mushrooms in "house cleaning," and in "preparing food," work in which they are still engaged.

FUNGI IN THE ARTS.

A number of different species of mushrooms have been employed in the manufacture of useful articles. Their use for such purposes, however, was more common in the past than at present, and it is largely therefore a matter of interest at the present time, though some are still employed for purposes of this kind.

Tinder mushroom, or amadou. — The *Polyporus fomentarius*, or "tinder mushroom" or, as it is sometimes called, "German tinder," was once employed in the manufacture of tinder. The outer hard coat was removed and the central portion, consisting almost entirely of the tube system of several years' growth, was cut into strips and beaten to a soft condition. In this form it was used as tinder for striking fire.

The inner portion was also used in making caps, chest-protectors, and similar articles. A process now in vogue in some parts of Germany, is to steam the fruit bodies, remove the outer crust, and then, by machinery constructed for the purpose, shave the fruit body into a long, thin strip by revolving it against a knife in much the same way that certain woods are shaved into thin strips for the manufacture of baskets, plates, etc. Some articles of clothing made from this fungus material are worn by peasants in certain parts of Europe.

Mushrooms for razor strops. — The beech polyporus (*P. betulinus*) several centuries ago was used for razor strops. The fruit body after being dried was cut into strips, glued upon a stretcher, and smoothed down with pumice stone (Asa Gray Bull. 7: 18, 1900). The sheets of the weeping merulius (see Fig. 189) were also employed for the same purpose, as were also the sheets of "punk" [Pg 235] formed from mycelium filling in cracks in old logs or between boards in lumber piles. Sometimes extensive sheets of this punk are found several feet long and a foot or more wide. These sheets of pure mycelium resemble soft chamois skin or soiled kid leather.

Mushrooms employed for flower pots. — In Bohemia (according to Cooke, Fungi, etc., p. 103) hoof-shaped fruit bodies of *Polyporus fomentarius* and *igniarius* are used for flower pots. The inner, or tube portion, is cut out. The hoof-shaped portion, then inverted and fastened to the side of a building or place of support, serves as a receptacle for soil in which plants are grown.

Curios. — The *Polyporus applanatus* is much sought by some persons as a "curio," and also for the purpose of etching. In the latter case they serve as pastels for a variety of art purposes. The under surface of the plant is white. All collectors of this plant know that to preserve the white fruiting surface in a perfect condition it must be handled very carefully. A touch or bruise, or contact with other objects mars the surface, since a bruise or a scratch results in a rapid change in color of the injured surface. Beautiful etchings can thus be made with a fine pointed instrument, the lines of color appearing as the instrument is drawn over the surface.

Fungi for medicinal purposes.—A number of the fungi were formerly employed in medicine for various purposes, but most of them have been discarded. Some of the plants were once used as a purgative, as in the case of the officinal polyporus, the great puff ball, etc. The internal portion of the great puff ball has been used as an anodyne, and "formidable surgical operations have been performed under its influence." It is frequently used as a narcotic. Some species are employed as drugs by the Chinese. The anthelmintic polyporus is employed in Burmah as a vermifuge. The ergot of rye is still employed to some extent in medicine, and the ripe puff balls are still used in some cases to stop bleeding of wounds.

Luminosity of fungi.—While the luminosity possessed by certain fungi cannot be said to be of distinct utility, their phosphorescence is a noteworthy phenomenon. That decaying wood often emits this phosphorescent light has been widely observed, especially in wooded districts. It is due to the presence of the mycelium of one of the wood destroying fungi. The luminosity is often so bright that when brought near a printed page in the dark, words can be read. Hawthorne "reported the light from an improvised torch of mycelium infected wood, to have carried him safely several miles through an otherwise impassable forest." (Asa Gray, Bull. 7: 7, 1900). The sulphur polyporus is said sometimes to be phosphorescent. The [Pg 236] *Clitocybe illudens* (see Fig. 92) has long been known to emit a strong phosphorescent light, and has been called "Jack-my-lantern." This plant often occurs in great abundance. At mountain hotels it is often brought in by day, and the guests at night, discovering its luminosity, trace grotesque figures, or monograms, on the ground by broken portions, which can be seen at a considerable distance. *Lentinus stipticus* in this country is also phosphorescent. In Europe, the *Pleurotus olearius* (very closely related to our *Clitocybe illudens*) on dead olive trunks is one of the best known of the phosphorescent species. Other phosphorescent species are, according to Tulasne, *A. igneus* from Amboyna, *A. noctileucus* in Manila, and *A. gardneri* in Brazil.

The use of certain mushrooms in making intoxicant beverages is referred to in Chapter XXII.

Since the artificial cultivation of mushrooms for food is becoming quite an industry in this country with some, the following chapter is devoted to a treatment of the subject. Mention may be made here, however, of the attempts in parts of France to cultivate truffles, species of subterranean fungi belonging to the ascomycetes (various species of the genus *Tuber*). It had long been observed that truffles grow in regions forested by certain trees, as the oak, beech, hornbeam, etc. Efforts were made to increase the production of truffles by planting certain regions to these trees. Especially in certain calcareous districts of France (see Cooke, Fungi, etc., p. 260) young plantations of oak, beech, or beech and fir, after the lapse of a few years, produced truffles. The spores of the truffles are in the soil, and the mycelium seems to maintain some symbiotic relation with the roots of the young trees, which results in the increase in the production of the fruit bodies. Dogs and pigs are employed in the collection of truffles from the ground.

Comparatively few of the truffles, or other subterranean fungi, have been found in America, owing probably to their subterranean habit, where they are not readily observed, and to the necessity of special search to find them. In California, however, Dr. Harkness (Proc. Calif. Acad. Sci.) has collected a large number of species and genera. Recently (Shear. Asa Gray Bull. 7: 118, 1899) reports finding a "truffle" (*Terfezia oligosperma* Tul.) in Maryland, and *T. leonis* occurs in Louisiana. [Pg 237]

FOOTNOTES:

[C] There is not room here to discuss the uses of other fungi than the "mushrooms."

CHAPTER XX.

CULTIVATION OF MUSHROOMS.

The increasing interest in mushrooms during the past few years has not been confined to the kinds growing spontaneously in fields and woods, but the interest aroused in the collection and study of the wild varieties has been the means of awakening a general interest in the cultivation of mushrooms. This is leading many persons to inquire concerning the methods of cultivation, especially those who wish to undertake the cultivation of these plants on a small scale, in cellars or cool basements, where they may be grown for their own consumption. At somewhat frequent intervals articles appear in the newspapers depicting the ease and certainty with which mushrooms can be grown, and the great profits that accrue to the cultivator of these plants. While the profits in some cases, at least in the past, have been very great to cultivators of mushrooms, the competition has become so general that through a large part of the year the market price of mushrooms is often not sufficient to much more than pay expenses. In fact, it is quite likely that in many cases of the house cultivation of mushrooms the profits are no larger, taking the season through, than they are from the cultivation of tomatoes or other hothouse vegetables. Occasionally some persons, who may be cultivating them upon a small scale in houses erected for some other purpose, or perhaps partly used for some other purpose, may succeed in growing quite a large crop from a small area with little expenditure of time and money. The profits figured from such a crop grown on a small scale where the investment in houses, heating apparatus, and time, is not counted, may appear to be very large, but they do not represent the true conditions of the industry where the expense of houses and the cost of time and labor are taken into consideration.

Probably the more profitable cultivation of mushrooms in this country is where the cultivation is practiced on quite a large scale, in tunnels, or caves, or abandoned mines, where no expense is necessary in the erection of houses. The temperature throughout the year is favorable for the growth of the mushrooms without artificial heating. It is possible, also, to grow them on a large scale during the warm summer months when it is impossible to grow [Pg 238] them under the present conditions in heating house structures, and also when the market price of the mushrooms is very high, and can be controlled largely by the grower. For this reason, if it were possible to construct a house with some practical system of cooling the air through the summer, and prevent the drip, the cultivation in houses would probably be more profitable.

Figure 223.—View in Akron "tunnel," N. Y. Mushroom Co. Beds beginning to bear. Copyright.

For the past few years the writer has been giving some attention to the different methods of the cultivation of mushrooms in America, and in response to the growing interest for information concerning the artificial cultivation of these plants, it has seemed well to add this chapter on the cultivation of mushrooms to the second edition of the present work. The cultivation as practiced in America exists under a great variety of conditions. All of these conditions have not been thoroughly investigated, and yet a sufficient number of them have been rather carefully studied to warrant the preparation of this chapter. The illustrations which have been made from time to time, by flash light, of the cave culture of mushrooms in America, as well as of the house culture, will serve to illustrate graphically some of the stages in the progress of the work. For present purposes [Pg 239] we will consider, first, the conditions under which the cultivation is carried on, followed by a discussion of the principles involved in the selection and preparation of the material, the selection and planting of the spawn, as well as the harvesting of the crop.

THE CAVE CULTURE OF MUSHROOMS IN AMERICA.

Figure 224.—View in Akron "tunnel," N. Y. Mushroom Co. Beds beginning to bear. Copyright.

This has been practiced for a number of years in different parts of the Eastern United States, but perhaps only a small portion of the available caves or tunnels are at present used for this purpose. These subterranean mushroom farms are usually established in some abandoned mine where, the rock having been removed, the space is readily adapted to this purpose, if portions of the mine are not wet from the dripping water. The most extensive one which I have visited is located at Akron, New York, and is operated by the New York Mushroom Company. In a single abandoned cement mine there are 12 to 15 acres of available space; about 3 to 5 acres of this area are used in the operations of the culture and handling of materials. The dry portions of the mine are selected, and flat beds are made upon the bottom rock, with the use of hemlock boards, [Pg 240] making the beds usually 16 feet long by 4 feet wide, the boards being 10 inches wide. In this case, the beds, after soiling or finishing, are 9 inches deep, the material resting directly upon the rock, the boards being used only to hold the material on the edges in position. Figures 223 and 224 illustrate the position of the beds and their relation to each other, as well as showing the general structural features of the mine. The pillars of rock are those which were left at the time of mining, as supports for the rock roof above, while additional wood props are used in places. In this mine all of the beds are constructed upon a single plan.

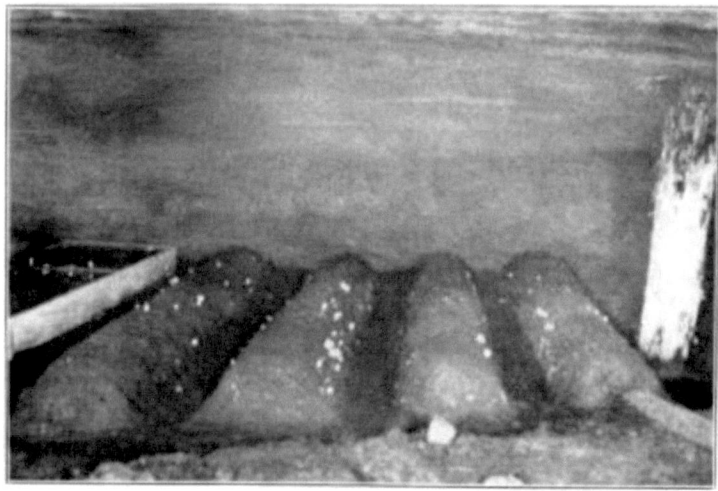

Figure 225.—View in Wheatland cave, showing ridge beds, and one flat bed. Copyright.

At another place, Wheatland, New York, where the Wheatland Cave Mushrooms are grown, beds of two different styles are used, the flat beds supported by boards as described in the previous case, and the ridge beds, where the material, without any lateral support, is arranged in parallel ridges as shown in Fig. 225. This is the method largely, if not wholly employed in the celebrated mushroom caves at Paris, and is also used in some cases in the outdoor cultivation of mushrooms. As to the advantage of one system of bed over the other, one must consider the conditions involved. Some believe a larger crop of mushrooms is obtained where there is an opportunity, as in the ridge beds, for the mushrooms to appear on the sides as well as on the upper surface of the beds. In the flat beds the mushrooms [Pg 241] can appear only at the upper surface, though occasionally single ones crop out in the crevice between the side board and the rock below.

Figure 226.—Single mushroom house (Wm. Swayne, Kennett Square, Pa.), "curing" shed at left. This house is heated in connection with other hothouses.

Probably at Paris, and perhaps also at some other places where the system of ridge beds is used, the question of the cost of the lumber is an important one, and the system of ridge beds avoids the expense of this item of lumber. In other cases, where the flat beds are used with the board supports, the cost of lumber is considered a small item when compared with the additional labor involved in making the ridge bed. The flat beds are very quickly made, and the material in some cases is not more than 7 inches deep, allowing a large surface area compared with the amount of food material, for the growth of the mushrooms. It may be possible, with the flat, shallow bed system, that as many or more mushrooms are obtained from the same amount of manure, as in the case of the ridge beds. When we consider the cost of the manure in some places, this item is one which is well worth considering.

THE HOUSE CULTURE OF MUSHROOMS.

Where this method of cultivation is employed, as the main issue, houses are constructed especially for the purpose. In general the houses are of two kinds. Those which are largely above the ground, and those where a greater or lesser pit is excavated so that the larger part of the house is below ground. Between these extremes all gradations exist. Probably it is easier to maintain an equable temperature when the house is largely below ground. Where it is largely above ground, however, the equability [Pg 242] of the temperature can be controlled to a certain extent by the structure of the house. In some cases a wall air space is maintained around the sides and also over the roof of the building. And in some cases even a double air space of a foot or 18 inches each is maintained over the roof. In some cases, instead of an air space, the space is filled with sawdust, single on the sides of the house, and also a 12 or 18-inch space over the roof. The sides of the house are often banked with earth, or the walls are built of stone or brick.

Figure 227.—Double mushroom house (L. S. Bigony's Mushroom Plant.) Packing room at left, "curing" shed at right, next to this is boiler room.

All of these houses, no matter what the type of construction, require ventilation. This is provided for by protected openings or exits through the roof. In some cases the ventilators are along the side of the roof, when there would be two rows of ventilators upon the single gable roof. In other cases a row of ventilators is placed at the peak, when a single row answers. These ventilators are provided with shut-offs, so that the ventilation can be controlled at will. The size of the house varies, of course, according to the extent of the operations which the grower has in mind.

The usual type of house is long and rather narrow, varying from 50 to 150 feet long by 18 to 21 or 24 feet wide. In some cases the single house is constructed upon these proportions, as shown by Fig. 226, with a gable roof. If it is desired to double the capacity of a house, two such houses are built parallel, the intercepting wall supporting the adjacent roof of the two houses, as shown in Fig. 227. A still further increase in the capacity of the house is often effected by increasing the number of these houses side by side. This results in a series of 8 or 10 houses forming one consolidated block of houses, [Pg 243] each with its independent ridge roof and system of ventilation. The separating walls between the several houses of such a block are probably maintained for the purpose of better controlling the temperature conditions and ventilation in various houses. If desired, communication from one house to another can be had by doors.

Interior structure and position of the beds. — The beds are usually arranged in tiers, one above the other, though in some houses the beds are confined only to the floor space. Where they are arranged in tiers in a house of the proportions given above, there are three tiers of beds. There is one tier on either side, and a tier through the middle; the middle tier, on account of the peak of the roof at this point, has one more bed than the tiers on the side. The number of beds in a tier will depend on the height of the house. Usually the house is constructed of a height which permits three beds in the side tier and four in the center tier, with an alley on either side of the center tier of beds, giving communication to all. If the house is very long and it is desirable, for convenience in passing from one house to another, to have cross alley-ways, they can be arranged, but the fewer cross alleys the larger surface area there is for beds.

Figure 228. — View in mushroom house (Wm. Swayne), showing upper bed in left hand tier. Copyright.

The size of the beds is governed by convenience in making the beds and handling the crop. The beds on the side tiers, therefore, are often three to three and one-half feet in width, affording a convenient [Pg 244] reaching distance from the alley. The beds of the center tier have access from the alley on either side and are usually

385

seven feet in width. The width of the alley varies according to the mind of the owner, from two to three or three and one-half feet. The narrow alley economizes space in the structure of a house; the wide alley, while slightly increasing the cost of the structure, makes it much more convenient in handling the material, and in moving about the house. The beds are constructed of one-inch boards. Various kinds of lumber are used, the hemlock spruce, the oak, Georgia pine, and so on. The beds are supported on framework constructed of upright scantling and cross stringers upon which the bottom boards are laid. These occur at intervals of three to four feet. The board on the side of each bed is 10 to 12 inches in width. The bottom bed, of course, is made on the ground. The upper beds in the tier are situated so that the distance is about three feet from the bottom of one bed to the bottom of the next above. Figs. 228 to 231 show the general structure of the beds.

Heating.—One portion of the house is set apart for the boiler room, where a small hot water heater is located. The position of the heater in one of these houses is shown in Fig. 227. In other cases, where the plant is quite a large one, a small separate or connecting boiler apartment is often constructed. In other cases, where the house is connected with or adjoining a system of greenhouses devoted to hothouse vegetables, the water pipes may run from the general boiler house which supplies the heat for all the houses. The water pipes in the mushroom houses are sometimes run beneath the boards or the walk in the alley, or in other cases are run just beneath the roof of the building.

Cultivation of mushrooms under benches in greenhouses.—This method is practiced to quite a large extent by some growers. In the house of Mr. William Swayne, Kennett Square, Pa., a number of large houses, devoted through the winter to the growing of carnations, are also used for the cultivation of mushrooms, a single long bed being made up underneath the beds of carnations. In these houses the water pipes providing heat for the building run along the sides of the building underneath the carnation beds at this point. Under these beds, where the water pipes run, no mushroom beds are made, since the heat would be too great, but under the three middle rows of beds in the house, mushroom beds are located. In this way, in a number of houses, several thousand square feet of surface for mushroom beds can be obtained. The carnations are grown, not in pots, but in a general bed on a bench. In watering the carnations, care [Pg 245] is used in the distribution of the water, and in the amount used, to prevent a surplus of water dripping through on the mushrooms below.

Cellar culture. — For the cultivation of mushrooms on a small scale, unoccupied portions of cellars in a dwelling house are often used. The question is sometimes asked if it is injurious to the health of the family in a dwelling house when mushrooms are grown in the cellar. Probably where the materials used in making up the beds are thoroughly cured before being taken into the cellar, no injurious results would come from the cultivation of the plant there. In case the manure is cured in the cellar, that is, is there carried through the process of heating and fermentation in preparation for the beds, the odors arising from the fermenting material are very disagreeable to say the least, and probably are not at all beneficial to one's general health.

Figure 229.—View in mushroom house (Wm. Swayne). View down alley on right hand side. Copyright.

In the cellar culture of mushrooms the places selected are along the sides of the cellar in unused portions. Floor beds alone may be made by using the boards to support one side, while the wall forms [Pg 246] the support on the other side as in the arrangement of beds on the side tiers in the mushroom houses; or tiers of beds may be arranged in the same way, one bed on the bottom, and one or two beds above. The number of beds will vary according to the available space. Sometimes, where it is not convenient to arrange the larger beds directly on the bottom of the cellar, or in tiers, boxes three or four feet, or larger, may be used in place of the beds. These can be put in out of the way places in the cellar. The use of boxes of this description would be very convenient in case it was desired to entirely do away with the possibility of odors during the fermentation of the manure, or in the making up of the bed. Even though the manure may be cured outside of the cellar, at the time it is made in the beds the odors released are sometimes considerable, and for several days might be annoying and disagreeable to the occupants of the dwelling, until such a time as the temperature of the manure had dropped to the point where the odors no longer were perceptible. In this case, with the use of boxes, the manure can be cured outside, made into beds in the boxes and taken into the cellar after the temperature is down to a point suitable for spawning, and very little odor will be released. If there is a furnace in the cellar it should be partitioned off from the portion devoted to mushroom culture.

Cultivation in sheds or out of the way places.—It is possible to grow mushrooms in a number of places not used for other purposes. In sheds where the beds may be well protected from the rain and from changing currents of air, they may be grown. In open sheds the beds could be covered with a board door, the sides of the bed being high enough to hold the door well above the mushrooms. In the basements of barns, or even in stables where room can be secured on one side for a bed, or tier of beds, they are often grown successfully.

Garden and field culture of mushrooms.—In Europe, in some cases, mushrooms are often grown in the garden, ridge beds being made up in the spring and spawned, and then covered with litter, or with some material similar to burlaps, to prevent the complete drying out of the surface of the beds. Sometimes they are cultivated along with garden crops. Field culture is also practiced to some extent. In the field culture rich and well drained pastures are selected, and spawned sometime during the month of May. The portions of spawn are inserted in the ground in little T-shaped openings made by two strokes of the spade. The spade is set into the ground once, lifted, and then inserted again so that this first slit is on one side of the middle of the spade and perpendicular to it. The spade is inserted [Pg 247] here and then bent backwards partly so as to lift open the sod in the letter T. In this opening the block of spawn is inserted, then closed by pressure with the foot. The spawn is planted in this way at distances of 6 to 8 feet. It runs through the summer, and then in the autumn a good crop often appears.

CURING THE MANURE.

Selection of manure.—Horse manure is the material which is most generally used, though sometimes a small percentage of other manures, as sheep manure, is added. In the selection of the manure it is desirable to obtain that which is as fresh as possible, which has not passed through the stage of fermentation, and which contains some straw, usually as litter, but not too large a percentage of straw. Where there is a very large percentage of straw the manure is usually shaken out with a fork, and the coarser portion removed. If there is not too much of this coarse material the latter is often cured in a separate pile and used for the bottom of the beds, the finer portions of the manure, which have been separated, are used for the finishing and for the bulk of the bed.

Figure 230.—View in mushroom house (L. S. Bigony). View on top of fourth bed, middle tier. Copyright.

Where manure is obtained on a large scale for the cultivation in houses or in caves, it is usually obtained by the carload from liveries in large cities. It is possible to contract for manure of certain livery [Pg 248] stables so that it may be obtained in a practically fresh condition, and handled by the liverymen according to directions, which will keep it in the best possible condition for the purpose. In the cave culture of mushrooms the manure is usually taken directly into the caves, and cured in some portion of the cave. In the house cultivation of mushrooms there is usually a shed constructed with an opening on one or two sides, at the end of the house connected with the beds, where the manure may be cured. In curing it, it is placed in piles, the size of which will depend upon the amount of manure to be cured, and upon the method employed by the operator. The usual size, where considerable manure is used, is about three feet in depth by ten or twelve feet wide, and fifteen to twenty feet long. The manure is laid in these piles to heat, and is changed or turned whenever desirable to prevent the temperature from rising too high. The object of turning is to prevent the burning of the material, which results at high degrees of temperature in fermentation. It is usually turned when the temperature rises to about 130° F. At each turning the outside portions are brought to the center of the pile. The process is continued until the manure is well fermented and the temperature does not rise above 100 to 120 degrees, and then it is ready for making into beds.

There are several methods used in the process of curing, and it does not seem necessary that any one method should be strictly adhered to. The most important things to be observed are to prevent the temperature from rising too high during the process of fermentation, to secure a thorough fermentation, and to prevent the material from drying out, or burning, or becoming too wet. The way in which the material is piled influences the rapidity of fermentation, or the increase of temperature. Where the material is rather loosely piled it ferments more rapidly, and the temperature rises quickly. Watering the manure tends to increase the rapidity of fermentation and the elevation of the temperature. It is necessary, though, sometimes to water the material if the heat has reached such a point that it is becoming too dry, or if there is a tendency for it to burn. The material is then turned, and watered some, but care should be used not to make it too wet, since the spawn will not run in wet material.

In general we might speak of three different methods in the curing of the manure. *First, the slow process of curing.* According to this method, which is practiced by some, the time of fermentation may extend from four to five weeks. In this case the manure is piled in such a way that the temperature does not rise rapidly. [Pg 249] During the four or five weeks the manure is turned four or five times. The turning occurs when the temperature has arisen to such a point as to require it.

Another method, used by some, might be called a rapid process of curing. According to this, the time for curing the manure extends over a period of about a week, or five to ten days. The material is piled in such a way as to cause rapid fermentation and rapid rising of temperature, the material sometimes requiring to be turned every day or two, sometimes twice a day, in order to lower the temperature and prevent the material from burning or drying out. Between this rapid process of curing, and the slow process of curing, the practice may extend so that, according to the method of different operators, the period of curing extends from one week to a month or five weeks.

Figure 231.—View in mushroom house (L. S. Bigony's Mushroom Plant, Lansdale, Pa.), showing alley and side tier of beds. Copyright.

The third method of curing consists in putting the material at once into the beds before curing, and mixing in with the manure, as it is placed in the bed, about one part of loam or garden soil to four or five parts of the fresh manure. The material is then left in this condition to cure without changing or turning, the temperature rising perhaps not above 130° F. With some experience in determining the firmness with which the bed should be made to prevent a too high rise of temperature, this practice might prove to be successful, [Pg 250] and would certainly save considerable labor and expense in the making of the beds. Mr. William Swayne of Kennett Square, Pa., in the winter of 1900–1901, made up a portion of one of his beds in this way, and no difference could be seen in the results of the crop, the crop from the beds made in this way being as good as that of the adjoining beds, and he intends the following year to make up all of his beds in the same way.

Mixing soil with the manure at the time of fermentation.—While in the cave culture of mushrooms the manure is usually fermented and used without the admixture of soil, usually in the house or cellar culture rich loam soil, or rotted sod, is mixed with the manure at the time of turning it, during the process of fermentation. At the time of the first turning, soil is mixed in, a layer of the manure being spread out on the ground, and then a sprinkling of soil over this. Then another layer of the manure is added with another sprinkling of soil, and so on as the new pile is built up. In the first turning of the manure, about one part of soil is used to eight or nine parts of manure. Then at the last turning another mixture of soil is added, so that there is about one-fifth part soil in the mixture. The soil aids somewhat in lowering the temperature, and also adds some to the bulk, so that more beds can be made up with the same amount of manure.

Horse droppings free from straw.—For growing mushrooms on a small scale, as in cellars or boxes, some prefer to select the horse droppings free from straw.

MAKING UP THE BEDS.

Making up beds without the addition of soil.—In the cave culture of mushrooms the beds are usually made from manure alone, there being no addition of soil. This is perhaps partly due to the expense of getting the soil in and out from the caves as well as to the low temperature prevailing there. It is believed by many that the results are equally as good in beds from the manure alone as in those which contain an admixture of soil. The method of making the beds in the Akron cave, or "tunnel," is as follows: The manure, immediately after it has passed through the process of fermentation and curing in the pile, is carted to the district in the mine where the beds are to be made and is dumped in a long windrow on the ground. The length of the windrow depends of course upon the amount of material which is ready, as well as upon the amount necessary for making up the beds for that distance. Two hemlock boards, sixteen feet long and ten inches wide, and two, four feet long and the [Pg 251] same width, are then hastily nailed into the form of a rectangular frame. This is placed upon the rock bottom at one end of the row of material, perpendicular to it usually.

Figure 232.—View in Akron "tunnel," N. Y. Mushroom Co. Making up the beds. Copyright.

The workmen then, with forks, distribute the material in this frame. If there is coarser material which has been separated from the finer material, this is placed in the bottom of the bed and the finer material is then filled on top. A layer of material is distributed over the bottom and then tamped down by striking with the back of the fork, as shown in Fig. 232. In this figure the material is shown to be off at one end of the bed. This was in a section of the mine where it was not convenient to follow the beds in the direction of the pile of manure, so that the material is distributed on from the end of the bed instead of from the side, as is the usual method. After several inches have been distributed in this way and tamped down with the back of the fork, the operator tramps over the material with his feet and presses it down more firmly. Another layer of material is distributed over this, and tamped and tramped down in a similar manner. The operation is repeated until the depth of the manure after tramping down is about seven inches. It is then left for the completion of the curing process and for the lowering of the temperature to the desired point. Usually, after making the bed in this way, there is a rise in the temperature for several days, gradually lowering until finally it reaches the point favorable for planting the spawn. [Pg 252]

Where the beds are made successively, one after another, following the windrow of manure, the material used for the first bed removes from the windrow a sufficient amount to make room for the second bed, and in like manner room for the successive beds is provided for as the material is taken for each one, so that the frames are put together and the beds are formed rapidly and easily.

Making ridge beds in caves. — In the making of the ridge beds in caves there are two methods which might be spoken of. One method is the well known one practiced in certain of the caves near Paris, where the material is taken by workmen in large baskets and distributed in rows. The ridge is gradually formed into shape by walking astride of it, as additional material is emptied on from the baskets, the workmen packing and shaping the ridge by pressure from their limbs as they stand astride of the row. In this way the ridges are made as high or somewhat higher than their breadth at the base, and quite near together, so that there is just room in many cases to walk between the beds. In one cave in America, where the ridge system is used to some extent, the ridges are made with the aid of a board frame the length of the bed and the width of the base of the ridge. The long boards of this frame are slanting so that they are more or less the shape of the ridge, but not equal to its height. This frame is placed on the rock bottom, filled with manure and tramped on by the workmen. Then the frame is lifted on the ridge and more material is added and tramped on in like manner, until the bulk of the ridge bed is built up in this way and compressed into shape.

Beds in Houses Constructed for the Purpose of Growing Mushrooms. — Where only the floor of the house is used, a middle bed and two side beds are sometimes formed in the same manner as described in the construction of the house for the tiers of beds, with an alley on either side of the large center bed, giving access to all. In some cases the entire surface of the bottom is covered with material, but divided into sections of large beds by framework of boards, but with no alleys between. Access to these beds is obtained by placing planks on the top of the boards which make the frame, thus forming walks directly over portions of the bed. In some cases ridge beds, as described for cave cultivation, are made on the floor of these houses. The beds are filled in the same way as described for the cave culture of mushrooms, but usually, in the beds made in houses built for the purpose of growing mushrooms, a percentage of soil is mixed in with the manure, the soil being usually mixed in at the time of turning the manure during the process of fermentation. [Pg 253] Garden soil or rich loam is added, say at the first time the manure is turned while it is fermenting. Then, some time later during the process of fermenting, another admixture of soil is added. The total amount of soil added is usually equal to about one-fifth of the bulk of the manure.

As this material, formed of the manure with an admixture of soil, is placed in the beds it is distributed much in the same manner as described for the making of flat beds in caves or tunnels. Usually, however, if there is coarse material which was separated from the manure at the first sorting, this without any mixture of soil is placed in the bottom of the bed, and then the manure and soil is used for the bulk of the bed above. This coarser material, however, is not always at hand, and in such cases the beds are built up from the bottom with the mixture of manure and soil. The depth of the material in the beds in these houses varies according to the experience of the operator. Some make the beds about eighteen inches in depth, while others do not make the beds more than eight or ten or twelve inches in depth. Where there are tiers of beds, that is, one bed above the other, very often the lowest bed, the one which rests directly upon the ground, is made deeper than the others.

While it is the general custom to use material consisting of an admixture of manure and soil in the proportions described, this custom is not always followed. In the case of the beds which are made up in the summer for the fall and early winter crop, soil, being easily obtained at that season of the year, is mixed with the manure. Some growers, however, in making the beds in midwinter for the spring crop, do not use any soil since it is more difficult to obtain it at that season. In such cases the beds are made up of manure alone. The experience in some cases shows that the crop resulting from this method is equally as good as that grown where soil has been added. In the experience of some other growers a bin of soil is collected during the summer or autumn which can be used in the winter for mixing in with the manure and making the beds for the spring crop. Where sod is used this is collected in pastures or fence rows in June, piled, and allowed to rot during the summer.

In distributing the material in the beds, the methods of packing it vary according to the wishes or experience of the grower. It is often recommended to pack the material very firmly. The feeling that this must be packed very thinly has led to the disuse of beds in tiers by some, because it is rather difficult to pack the material down very firmly where one bed lies so closely above another. Where the practice is followed of packing the material very firmly in [Pg 254] the bed, some instrument in the form of a maul is used to tamp it down. Where there are tiers of beds an instrument of this kind cannot well be used. Here a brick or a similar heavy and small instrument is used in the hand, and the bed is thus pounded down firmly. This is a tedious and laborious operation. Many growers do not regard it as essential that the beds should be very firmly packed. In such cases the material is distributed on the beds and the successive layers are tamped down as firmly as can well be done with the back of a fork or an ordinary potato digger, which can be wielded with the two hands in between the beds. In the experience of these growers the results seem to be just as good as where the beds are more firmly packed down.

It is the practice in some cases where the bed lies against the side of the house to build up the material of the bed at the rear, that is, at the side of the house, much deeper than at the front, so that the depth of the bed at the back may be eighteen to twenty inches or two feet, while the front is eight to ten or twelve inches. This provides a slightly increased surface because of the obliquity of the upper surface of the bed, but it consumes probably a greater amount of material. It probably is not advantageous where the operations are carried on on a large scale, where abundant room is available, where the material for making the beds is expensive, and it is desirable to obtain from the material all that can be drawn in a single crop. The same practice is sometimes recommended and followed in the case of the beds made in cellars.

In the making of beds with fresh material, that is, with unfermented manure, as was done by Mr. William Swayne of Kennett Square, Pa., one season, the coarser material is put in the bottom of the bed, and then as the manure is distributed in the bed the soil is sprinkled on also, so that finally when the bed is completed the proportions of soil and manure are the same as when it is mixed in at the time of fermentation. In making the beds in this way, should any one be led to attempt it, it would be necessary to guard against a too high temperature in the fermentation of this fresh material; the temperature should not run above 130 degrees. It would also require a longer time from the making of the bed to planting the spawn than in the case of those beds where the manure is fermented and cured before being made up. Probably the total amount of time from the beginning to the completion of the preparation of the bed for spawning would not be greater, if it would be so great.

The beds all having been made, they are left until they are in a suitable condition for spawning. The determination of this point, [Pg 255] that is, the point when the beds are ready for planting the spawn, seems to be one of the most important and critical features of the business. The material must be of a suitable temperature, preferably not above 90° F., and not below 70°. The most favorable temperature, according to some, other conditions being congenial, ranges from 80° to 85° F., while many prefer to spawn at 70° to 75°. Many of the very successful growers, however, do not lay so much stress upon the temperature of the bed for the time of spawning as they do upon the ripeness, or the cured condition, of the material in the bed. This is a matter which it is very difficult to describe to one not familiar with the subject, and it is one which it is very difficult to properly appreciate unless one has learned it by experience. Some judge more by the odor, or the "smell," as they say, of the manure. It must have lost the fresh manure "smell," or the "sour smell," and possess, as they say, a "sweet smell." Sometimes the odor is something like that of manure when spawn has partly run through it. It sometimes has a sweetish smell, or a smell suggestive of mushrooms even when no spawn has run through it.

Another important condition of the material is its state of dryness or moisture. It must not be too dry or the spawn will not run. In such cases there is not a sufficient amount of moisture to provide the water necessary for the growth of the mycelium. On the other hand, it must not be too wet, especially at the time of spawning and for a few weeks after. Some test the material for moisture in this way. Take a handful of the material and squeeze it. If on releasing the hold it falls to pieces, it is too dry. By squeezing a handful near the ear, if there is an indication of running water, even though no water may be expressed from the material, it is too wet. If on pressure of the material there is not that sense of the movement of water in it on holding it to the ear, and if on releasing the pressure of the hand the material remains in the form into which it has been squeezed, or expands slightly, it is considered to be in a proper condition so far as moisture is concerned for planting the spawn.

WHAT SPAWN IS.

The spawn of the mushroom is the popular word used in speaking of the mycelium of the mushroom. The term is commonly used in a commercial sense of material in which the mycelium is growing. This material is horse manure, or a mixture of one or two kinds of manure with some soil, and with the threads of the mycelium growing in it. The mycelium, as is well known, is the growing [Pg 256] or vegetative part of the mushroom. Sometimes the word "fiber" is used by the mushroom growers in referring to the mycelium which appears in the spawn, or in the mushroom bed. The mycelium is that portion of the plant which, in the case of the wild varieties, grows in the soil, or in the leaf mold, in the tree trunk or other material from which the mushroom derives its food. The threads of mycelium, as we know, first originated from the spore of the mushroom. The spore germinates and produces delicate threads, which branch and increase by growth in extent, and form the mycelium. So the term spawn is rarely applied to the pure mycelium, but is applied to the substratum or material in which spawn is growing; that is, the substratum and mycelium together constitute the spawn.

Natural spawn or virgin spawn. — This is termed natural spawn because it occurs under natural conditions of environment. The original natural spawn was to be found in the fields. In the early history of mushroom culture the spawn from the pastures and meadows where mushrooms grew was one of the sources of the spawn used in planting. The earth containing the spawn underneath clumps of mushrooms was collected and used.

It occurs more abundantly, however, in piles of horse manure which have stood for some time in barn yards, or very often in stalls where the manure is allowed to accumulate, has been thoroughly tramped down and then has been left in this condition for some time. It occurs also in composts, hothouse beds, or wherever accumulations of horse manure are likely to occur, if other conditions are congenial. The origin of the natural spawn under these conditions of environment is probably accounted for in many cases by the presence of the spores which have been in the food eaten by the horse, have passed through the alimentary canal and are thus distributed through the dung.

The spores present in the food of the horse may be due to various conditions. Horses which go out to pasture are likely to take in with the food obtained in grazing the spores scattered around on the grass, and in the upper part of the sod, coming from mushrooms which grew in the field. In other cases, the spores may be present in the hay, having been carried by the wind from adjacent fields, if not from those which have grown in the meadow. In like manner they may be present in the oats which have been fed to the horse. In the case of stable-fed animals, the inoculation of the manure in this way may not always be certain or very free. But in the case of pasture-fed horses which are stalled at night probably [Pg 257] the inoculation is very certain and very abundant, so that a large number of spores would be present in the manure from horses fed in this way.

The natural spawn also may originate from spores which are carried by the wind from the pasture or meadow mushrooms upon manure piles, or especially from spores which may lodge in the dust of the highways or street. Many of these spores would cling to the hoofs of the horses and at night, or at times of feeding, would be left with the manure in the stall. At other times horse droppings may be gathered from roads or streets where spores may be present in the dust. The piles of the droppings accumulated in this way, if left a sufficient time, may provide natural spawn by this accidental inoculation from the spores.

Probably few attempts have been made to grow the natural spawn with certainty in this country, though it does not appear to be an impracticable thing to do, since formerly this was one source of the virgin spawn in Europe. It is usually obtained by search through stables and barn yards or other places where piles of horse manure have accumulated and have remained for several months. In some cases the growers keep men employed through the summer season searching the yards and stables over a considerable area for the purpose of finding and gathering this natural spawn. It is probably termed virgin spawn because of its origin under these natural conditions, and never having been propagated artificially.

The natural spawn, as indicated above, is employed for a variety of purposes. It is used for inoculating the bricks in the manufacture of brick spawn. It is used for propagating once or twice in the mushroom beds, for the purpose of multiplying it, either in the manufacture of brick spawn, or for flake spawn, which is planted directly in the beds to be used for the crop. In some places in America it is collected on a large scale and relied on as the chief source of spawn for planting beds. In such cases the natural or virgin spawn is used directly and is of the first and most vigorous generation. It is believed by growers who employ it in this way that the results in the quality and quantity of the crop exceed those produced from the market spawn. But even these growers would not always depend on the natural spawn, for the reason, that collecting it under these conditions, the quantity is certain to vary from year to year. This is due probably to varying conditions of the season and also to the varying conditions which bring about the chance inoculation, or the accumulation of the material in the yard for a sufficient amount of time to provide the mycelium. [Pg 258]

It would be interesting, and it might also prove to be profitable to growers, if some attempt were made to grow natural spawn under conditions which would perhaps more certainly produce a supply. This might be attempted in several different ways. Stall-fed horses might be fed a ripe mushroom every day or two. Or from the cap of ripe mushrooms the spores might be caught, then mixed with oats and fed to the horse. Again, the manure piles might be inoculated by spores caught from a number of mushrooms. Manure might also be collected during the summer months from the highways and aside from the probable natural inoculation which this material would probably have from the spores blown from the meadow and pasture mushrooms, additional inoculation might be made. The manure obtained in this way could be piled under sheds, packed down thoroughly, and not allowed to heat above 100° F. These piles could then be left for several months, care being used that the material should have the proper moisture content, not too dry nor too wet. This is given only as a suggestion and it is hoped that some practical grower will test it upon a small scale. In all cases the temperature should be kept low during the fermentation of these piles, else the spawn will be killed.

One of the methods of obtaining natural spawn recommended by Cuthill ("Treatise on the Cultivation of the Mushroom") is to collect horse droppings all along the highways during the summer, mixing it with some road sand and piling it in a dry shed. Here it is packed down firmly to prevent the heat rising too high. A "trial" stick is kept in the pile. When this is pulled out, if it is so hot as to "burn the hand," the heat is too great and would kill the spawn. In several months an abundance of the spawn is generated here.

Mill-track spawn. — "Mill-track" spawn originated from the spawn found in covered roadways at mills or along tram-car tracks where horses were used. The accumulation of manure trodden down in these places and sometimes mixed with sawdust or earth, provided a congenial place for the growth of the mycelium. The spawn was likely introduced here through spores taken in with the food of the horse, or brought there from highways, if they were not already in the soil from mushrooms grown there. It would be then multiplied by the growth of the spawn, and from spores of mushrooms which might appear and ripen. The well tramped material in which the mycelium grew here, when broken up, formed convenient blocks of spawn for storage and transportation, and probably led to the manufacture of brick spawn.

Manufactured spawn. — The manufactured spawn, on the other [Pg 259] hand, is that which is propagated artificially by the special preparation of the substratum or material in which the mycelium is to grow. This material is inoculated either with a piece of natural spawn, or with pieces of previously manufactured spawn. It is put upon the market in two different forms; the brick spawn, and the flake spawn. The latter is sometimes known as the French spawn, while the former, being largely manufactured in England, is sometimes spoken of as the English spawn.

Figure 233.—Brick spawn. Three "bricks," one marked to show into how many pieces one brick may be broken.

Brick spawn.—The brick spawn is so called because the material in which the mycelium is present is in the form of bricks. These bricks are about 5 by 8 inches by 1-1/2 inches in thickness, and weigh about 1-1/4 pounds each when dried. The proportions of different kinds of material used in the manufacture of brick spawn probably vary with different manufacturers, since there is a difference in the size and texture of bricks from different sources. One method of making the brick spawn is as follows: Equal parts of horse dung, and cow dung, and loam soil are thoroughly mixed together to a consistency of mortar. This is pressed into the form of bricks and stood on edge to dry. When partly dry, a piece of spawn about an inch in diameter is pressed into one side of each brick. The bricks are then stood up again until thoroughly dried. They are then piled upon a layer of fresh horse manure about 8 inches deep, the pile of bricks being about 3 feet high. This pile is then covered over loosely with fresh horse manure, a sufficient amount to produce, [Pg 260] when heating, a temperature of about 100° F. They are left in this condition until the mycelium or "fiber" has thoroughly permeated the bricks. The spawn is now completed, and the bricks are allowed to dry. In this condition they are put upon the market. The bricks made with a very high percentage of soil often have the appearance of dried soil, with a slight admixture of vegetable matter.

Brick spawn from other sources presents a very different texture and contains probably a much larger percentage of horse manure, or, at least, a much smaller percentage of soil. The appearance of the brick is not that of soil with a slight admixture of vegetable materials, but has much the appearance of a dried and compressed mixture of horse dung and cow dung, with an abundance of the "fiber" or mycelium, "the greyish moldy, or thready matter," which constitutes the vital part of the spawn. In the selection of spawn this is an important item, that is, the presence of an abundance of "fiber" or mycelium. It can be seen on the surface, usually showing an abundance of these whitish threads or sheets, or a distinct moldy appearance is presented. On breaking the brick the great abundance of the "fiber" or whitish mycelium is seen all through it. This indicates that the brick possesses a high percentage of the "fiber," an important part of the spawn.

One not accustomed to the quality of spawn can therefore judge to a certain extent by the appearance of the bricks as to the quality, at least they can judge as to the presence of an abundance or a scanty quantity of the "fiber." Since the spawn remains in good condition for several years, there is usually no danger in the use of spawn which may be one or two years old. But it does deteriorate to some extent with age, and young spawn is therefore to be preferred to old spawn, provided the other desirable qualities are equal. Those who attempt to cultivate mushrooms, and depend on commercial or manufactured spawn, should see to it that the spawn purchased possesses these desirable qualities of texture, and the presence of an abundance of the mycelium. That which appears devoid of an abundance of mycelium should be rejected, and good spawn should be called for. There is no more reason why a grower should accept a worthless spawn from his seedsman than that he should accept "addled" eggs from his grocer. In this business, that is, the manufacture and sale of spawn, poor material is apt to be thrown on the market just as in the case of seeds, poor material may find its way upon the market. Sometimes this occurs through unscrupulous dealers, at other times through their ignorance, or through their failure to know the quality of the product they are handling. [Pg 261]

There are some brands of spawn, that is, those manufactured by certain houses, which rank very high among those who know the qualities and the value of good spawn. Some large growers send direct to the manufacturer for their spawn, and where it is to be obtained in large quantities this is a desirable thing to do, since the cost is much less. Where obtained from seedsmen in large quantities, the prices are much lower than where small quantities are purchased. One of these brands of spawn, the Barter spawn, is for sale by several different dealers, by Mr. H. E. Hicks, Kennett Square, Pa., by Henry F. Michell, 1018 Market street, Philadelphia, and by Henry Dreer, 724 Chestnut street, Philadelphia. Another brick spawn, known as "Watson Prolific," is for sale by George C. Watson, Juniper and Walnut streets, Philadelphia. James Vicks Sons, Rochester, N. Y., and Peter Henderson & Co., New York City, have their spawn manufactured expressly for their trade.

The Barter spawn is said to be made fresh every year, or every other year. Instead of the "continued culture" of spawn, that is, inoculating the bricks each succeeding year from the same line of spawn, which is, as it were, used over and over again, a return is made each year, or in the alternate years, to the natural or virgin spawn, which is obtained from old manure heaps. In this way, the Barter spawn [D] is within two to three, or four, generations of the natural spawn. The number of generations distant the brick is from the natural spawn, depends upon the number of times it may have been multiplied before it is inoculated into the bricks. That is, the natural spawn is probably first grown in large beds in order to multiply, to produce a sufficiently large quantity for the inoculation of the immense number of bricks to be manufactured. For it is likely that a sufficient amount of natural spawn could not be obtained to inoculate all the bricks manufactured in one year. If a sufficient amount of the natural or virgin spawn could be obtained to inoculate all the bricks of one year's manufacture, this would produce a spawn removed only one generation from that of natural spawn.

If the natural spawn were first grown in beds, and from here inoculated into bricks, this particular brick spawn would be removed two generations from the natural spawn. So the number of times that successive inoculations are made to multiply the spawn, the manufactured products are removed that many generations from the natural spawn. Where recourse is had to the natural, or virgin [Pg 262] spawn only once in two years, the second year's product would then be further removed from the natural spawn than the first year's product. Where we know that it is removed but one or a few generations from the natural spawn, it is a more desirable kind. For the nearer it is to the natural spawn, other things being equal, the more vigorous the mycelium, and the finer will be the mushrooms produced.

The brick spawn is sometimes manufactured in this country by growers for their own use, but at present it is manufactured on such a large scale in England that little or no saving is effected by an attempt to manufacture one's own brick spawn in this country.

Flake Spawn.—The flake spawn, or "flakes," is commonly known as the French spawn, because it is so extensively manufactured in France. It is made by breaking down beds through which the mycelium has run, and before the crop of mushrooms appears. That is, the bed is spawned in the ordinary way. When the mycelium has thoroughly permeated the bed, it is taken down and broken into irregular pieces, six to eight inches in diameter. Thus, the French spawn, where the beds are made entirely of horse manure, with no admixture of soil, consist merely of the fermented and cured manure, through which the mycelium has run, the material, of course, being thoroughly dried. This spawn may be removed one or several generations from the natural spawn.

Figure 234.—French spawn, or "flakes," ready to plant.

The French growers depend on natural spawn much more than American growers do. The natural spawn is collected from old [Pg 263] manure heaps. Beds made up in the ordinary way for the cultivation of mushrooms are planted with this. The mycelium is allowed to run until it has thoroughly permeated the manure. These beds are broken down and used to spawn the beds for the crop. In this case the crop would be grown from spawn only one generation removed from the virgin spawn. If a sufficient amount of natural spawn could not be obtained, to provide the amount required one generation old, it might be run through the second generation before being used. From the appearance of any spawn, of course, the purchaser cannot tell how many generations it is removed from the natural spawn. For this quality of the spawn one must depend upon the knowledge which we may have of the methods practiced by the different producers of spawn, if it is possible even to determine this.

SPAWNING THE BEDS.

The beds for growing the mushrooms having been made up, the spawn having been selected, the beds are ready for planting whenever the temperature has been sufficiently reduced and the material is properly cured. It is quite easy to determine the temperature of the beds, but it is a more difficult problem for the inexperienced to determine the best stage in the curing of the material for the reception of the spawn. Some growers rely more on the state of curing of the manure than they do upon the temperature. They would prefer to spawn it at quite a low temperature, rather than to spawn at what is usually considered an optimum temperature, if the material is not properly cured. The temperature at which different treatises and growers recommend that the bed should be spawned varies from 70° to 90° F. Ninety degrees F. is considered by many rather high, while 70° F. is considered by others to be rather low; 80° to 85° is considered by many to be the most favorable temperature, provided of course the other conditions of the bed are congenial. But some, so far as temperature is concerned, would prefer to spawn the bed at 75° F. rather than at 90°, while many recommend spawning at 70° to 75°. In some cases, I have known the growers to allow the temperature of the beds to fall as low as 60° before spawning, because the material was not, until that time, at the proper state of curing. Yet an experienced grower, who understands the kind of spawn to plant in such a bed, can allow the temperature to go down to 60° without any very great risk. Fresh spawn in an active state, that is, spawn which is in a growing condition, as may be obtained by tearing up a bed, or a portion of one, [Pg 264] through which the spawn has run, is better to plant in a bed of such low temperature. Or, a bed of such low temperature, after spawning, might be "warmed up," by piling fresh horse manure over it loosely for a week or ten days, sufficient to raise the temperature to 80° or 90°.

Figure 235.—Pieces of brick spawn ready to plant.

When the brick spawn is used, the method of planting varies, of course, with the methods of different operators. Some break the bricks into the desired size and plant the pieces directly in the bed, without any special preparation. The brick is broken into pieces about two or three inches in diameter. Some recommend breaking the brick of the ordinary size into about twelve pieces, some into nine pieces, so the custom varies with different operators. These pieces are planted from seven to nine inches apart in the bed. For example, if they are to be planted nine inches apart in the bed, holes are made, either with the hand or with some instrument, by pressing the material to one side sufficiently to admit of the piece of spawn being pressed in tightly. These openings are made, say, the first row on one side of the bed, about four and one-half inches from the side, and nine inches apart in the row. The second row is made nine inches from the first row, and so on. The pieces of spawn are inserted in the opening in the bed, and at a slight distance, two to three inches, below the surface. Some, however, insert the piece of spawn just at the level of the bed, the opening being such that the piece of spawn pressed into the opening is crowded below in place, and the surrounding material fits snugly on the sides. Thus, when the bed is spawned, the pieces may be a slight distance below the top of the bed when they can be covered by some material, or in [Pg 265] other cases, where the operator varies the method, they would lie just at the surface of the bed.

The bed is now firmed down according to the custom of the operator, either tamped down with some instrument very firmly, or by others, with the back of the fork or other similar instrument, the bed is made firm, but not quite so hard. The object in firming it down after spawning is to make the surface of the bed level, and also to bring the material in the bed very closely in touch on all sides with the spawn with which it is impregnated.

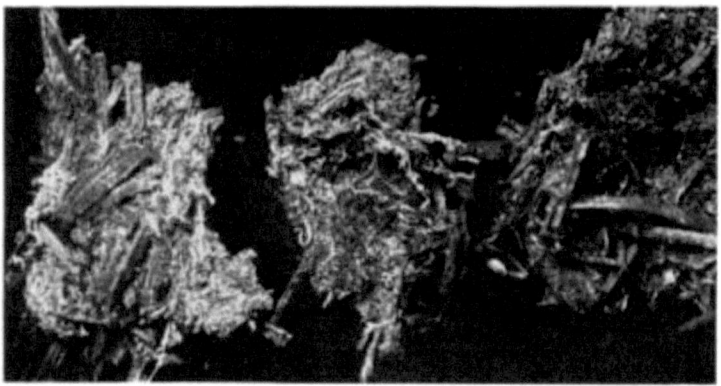

Figure 236.

- Piece of Natural Spawn.
- Piece of French Spawn.
- "Flakes" many generations old, "running out."]

Some growers follow the method of giving the spawn some little preparation before putting it into the bed. This preparation varies with different operators. Its object, however, is to slightly moisten the dry spawn, and perhaps, also, to very slightly start the growth. To accomplish this, some will cover the bricks, before breaking them, with fresh horse manure, and allow this to remain several days, so that the warmth and moisture generated here penetrate the material and soften somewhat the brick. Some pile it in a room or compartment where there is little moisture, until the bricks are permeated to some extent with the moisture, so that they are a little easier broken. They should not, under any circumstances, be wet or soft in the sense of having absorbed an excess of water, nor should they be stored for any length of time where they will be damp. Still others break the bricks into the desired pieces and place these directly on the top of the bed, at the place where they [Pg 266] wish to plant the piece of spawn. They are left here for two or three days on the surface of the beds. These pieces absorb some moisture and take up some warmth from the bed. Then they are planted in the ordinary way.

410

Spawning with Flake Spawn, or Natural Spawn.—In the use of the flake or natural spawn, the planting is accomplished in a similar way, but larger pieces of the spawn are used, two or three times the size of the pieces of brick employed. Some use a large handful. In some few cases, the growers use a flake spawn from their own crop. That is, each year a few beds are spawned from material which has been kept over from the previous season. This is often kept in boxes, in cool places, where it does not thoroughly dry out. In this way, the spawn is used over and over again, until it becomes much less vigorous than natural spawn, or a spawn which is only one or only a few generations distant from the natural spawn. This is seen in the less certainty with which the spawn runs through the bed, in the smaller crop of mushrooms, and their gradual deterioration in size. Some few practice the method of breaking down the bed after the crop has been nearly gathered, using this weak spawn to inoculate fresh beds. This practice is objectionable for the same reason that long cultivated spawn is objectionable.

Soiling the Beds.—After the beds have been planted with the spawn, the next thing is to soil them. That is, the manure in the bed is covered with a layer of loam soil, or garden soil, to the depth of two inches, then spread evenly over the bed, leveled off, and tamped down, though not packed too hard, and the surface is smoothed off. The time at which the soiling is done, varies also with different operators. Some soil immediately after planting the spawn. Others believe that the spawn will most certainly fail to run if the beds are soiled immediately after planting. These operators wait two or three weeks after the spawn has been planted to soil it. Others wait until the temperature of the bed has fallen from 80° or 85° at the time of spawning, to 70° or 60° F. Soiling at this temperature, that is, at 60° or 70° F., probably prevents the rapid cooling down of the bed, and it is desirable to soil, at least at this temperature, for that purpose. When the beds are soiled, they are then left until the crop is ready to gather. Some operators give no further attention to the beds after soiling, other than to water the beds, if that becomes necessary. It is desirable to avoid watering, if the bed can be kept at the right state of moisture without. In watering the beds while the spawn is running, there is danger of killing the young spawn with the water. Wherever it is necessary, [Pg 267] however, if the material in the

411

bed becomes too dry, lukewarm water should be used, and it should be applied through a fine rose of a watering pot.

While some operators after soiling the bed give no further care to it until the bed is bearing, others cover the beds with some litter, in the form of straw or excelsior. This is done for the purpose of conserving the moisture in the bed, and especially the moisture on the surface of the bed. Sometimes where there is a tendency for the material in the bed to become too dry, this litter on the surface retards the loss of moisture. Also, the litter itself may be moistened and the bed can absorb some moisture in this way, if it is desirable to increase the moisture content of the bed slightly.

When the spawn has once run well through the bed, watering can be accomplished with less danger of injury, yet great care must be used even now. The spawn will run through a bed with a somewhat less moisture content in the material than is necessary for drawing off the crop of mushrooms, though, of course, the spawn will not run if the bed is too dry. The only way to see if the spawn has run satisfactorily is to open up the bed at one or two points to examine the material, opening it up slightly. If the spawn has run well, a very delicate white "fiber," the mycelium, can be seen penetrating all through the material. This handful can be replaced in the bed, packed down, and the soil covered over and firmed again at this point.

When the mushrooms begin to appear, if the bed is a little dry, it should be watered from time to time through the fine rose of a watering pot. Lukewarm water should be used. Nearly all growers water the beds during the picking of the crop, or during the period of gathering the crop. At the first few waterings, water should not be sprinkled on the beds to wet them entirely through. Enough water is applied to diffuse a short distance only through the upper surface of the bed. At the next watering, several days later, the moisture is carried further down in the bed, and so on, through the several weeks, or months, over which the harvesting season extends. The object of thus gradually moistening the bed from above, is to draw the crop from the spawn at the surface of the bed first, and then, as the moisture extends downward, to gradually bring on the crop from the "fiber" below.

Gathering the Mushrooms. — In artificial cultivation, the mushrooms usually formed are very near, or on, the surface of the bed. In the case of the meadow or pasture mushrooms, they are formed further below the surface. This is probably due to the fact that the conditions [Pg 268] under which the mushrooms grow in cultivation are such that the surface of the bed is more moist, and is less subject to variations in the content of moisture, than is the surface of the ground in pastures. Although there may be abundant rains in the fields, the currents of air over the surface of the ground, at other times, quickly dries out the upper layers of the soil. But indoors the mycelium often runs to the surface of the bed, and there forms the numerous pinheads which are the beginnings of the mushrooms. The beds at this stage often present numerous clusters of the mycelium and these minute pinheads crowded very closely together. Hundreds or perhaps thousands of these minute beginnings of mushrooms occur within a small space. There are very few of these, however, that reach the point of the mature mushroom. Few only of the pinheads grow to form the button, and the others abort, or cease to grow. Others are torn out while the larger ones are being picked.

The time at which the mushrooms are picked varies within certain limits, with the different growers. Most cultivators, especially those who grow the mushrooms in houses, consider 60° F. the desirable temperature for the growth of mushrooms, that is, at a room temperature of 60° (while some recommend 57°). The temperature of the beds themselves will be slightly above this. Under these conditions, that is, where the mushrooms are grown at a room temperature of about 60°, they open very quickly. It is necessary here to gather the mushrooms before they open, that is, before the veil on the under surface breaks to expose the gill surface. This practice is followed, of course, within certain limits. It is not possible in all cases, to pick every mushroom before the veil breaks. They are collected once a day usually. At the time of collection all are taken which are of suitable size. Many of them may not yet have opened. But in the case of some of the older or more rapidly growing ones, the veil may have broken, although they have not expanded very much.

Some follow the method of having the fireman, on his round at night, when he looks after the fires in the heating room, gather the mushrooms. He passes through all parts of the house and picks the mushrooms which are of suitable size. These are gathered by grasping a single mushroom by the cap, or where there is a cluster of mushrooms close together, several are taken in the hand. The plant is twisted slightly to free the stem from the soil, without tearing it up to any great extent. They are thrown in this condition into baskets. The collector then takes them to the packing room, [Pg 269] and the following morning the plants are trimmed, that is, the part of the stems to which the earth is attached is cut away, the plants are weighed, put in baskets, and prepared for the markets. In other cases, the mushrooms are gathered early in the morning, in the same way, taken to the packing room, where the lower part of the stem is cut away, the plants are weighed, placed into the baskets and shipped to market.

Figure 237.—View in Packing Room (H. E. Hicks' Mushroom House, Kennett Square, Pa.) Copyright.

In some of the caves, or abandoned mines, which I have visited, where the mushrooms are grown on a large scale, the practice in picking the mushrooms varies somewhat from that just described.

414

In the first place, the mushrooms are allowed to stand on the bed longer, before they are picked. They are rarely, if ever, picked before they open. Mushrooms may be quite large, but if they have not opened, they are not picked. Very frequently, the plant may open, but, the operator says, it is not open enough. It will grow more yet. The object of the grower, in this case, is to allow the mushrooms to grow as long as it is possible, before picking, for the larger the mushroom, the more water it will take from the bed, and the more it weighs. This may seem an unprofessional thing for a grower to do, and yet it must be remembered that a large water content of the mushroom is necessary. The mushrooms grown in these mines are very firm and solid, qualities which are desired, not [Pg 270] only by the consumer, but are desirable for shipment. These mushrooms are much thicker through the center of the cap than those usually grown in houses at a room temperature of 60° F. For this reason, the mushrooms in these caves spread out more, and the edges do not turn up so soon. Since the cap is so thick and firm at the center, it continues to grow and expand for some little time after having opened, without turning up on the edges, and without becoming black and unsightly underneath. These large and firm mushrooms are not only desirable for their shipping qualities, but also, if they are not too large, they are prized because they are of such a nice size for broiling.

It is quite likely that one of the important conditions in producing mushrooms of this character is the low temperature of the mine. The temperature here, in July and August, rises not higher than 58° F., that is, the room temperature of the mines; while in the winter it falls not lower than 52°. The growth of mushrooms, under these conditions, may not be quite so rapid as in a house maintaining a room temperature of 60°. The operator may not be able to grow so many crops from the same area, during the same length of time; but the very fact that this low temperature condition retards the growth of the mushrooms is perhaps an important item in producing the firm and more marketable product, which can be allowed to grow longer before it is picked. It is possible, also, that another condition has something to do with the firmness and other desirable qualities of these mushrooms. It is, perhaps, to be found in the fact that natural spawn is largely used in planting the beds, so that the spawn is more vigorous than that which is ordinarily used in planting, which is several or many generations distant from the virgin condition.

The methods of picking in this mine differ, also, from those usually employed by growers of mushrooms. The mushrooms are pulled from the bed in the same way, but the operator carries with him two baskets and a knife. As fast as the mushrooms are pulled, and while they are still in hand, before the dirt can sift upon the other mushrooms, or fall in upon the gills of those which are open, the lower part of the stem is cut off. This stem end is then placed in one basket, while the mushrooms which have been trimmed are placed in another basket. In cutting off the stems, just enough is cut to remove the soil, so that the length of the stem of the mushroom varies. The mushrooms are then taken to the packing room in the cleanest possible condition, with no soil scattering therefrom or falling down among the gills, as occurs to a greater or lesser extent [Pg 271] where the mushrooms are picked and thrown indiscriminately into baskets.

Packing the Mushrooms. — In the packing room the mushrooms are prepared for shipment to market. The method at present usually employed is to ship them in baskets. The baskets vary in size, according to the market to which the mushrooms are to be shipped. They hold from three, to four, five, six, or ten pounds each. The larger baskets are only used where the mushrooms are shipped directly to the consumers. When the customer requires a large number of mushrooms, they can be shipped in these larger baskets. Where they are shipped to commission merchants, and the final market is not known to the packer, they are usually packed in small baskets, three to four or five pounds. The baskets are sometimes lined with paper; that is, at the time of the packing the paper is placed in the basket, one or two thicknesses of paper. The number of layers of paper depends somewhat upon the conditions of transportation. The greater amount of paper affords some protection from cold, in cold weather, and some protection from the evaporation of the moisture, in dry weather. When the basket is filled with the required quantity of mushrooms, which is usually determined first by weight, the surplus paper is folded over them. This is covered in most cases by thin board strips, which are provided for basket shipment of vegetables of this kind. In some cases, however, where shipped directly to customers so that the baskets soon reach their destination, additional heavy paper, instead of the board, may be placed over and around the larger part of the basket, and then tied down neatly with cord.

Placing the Mushrooms in the Basket.—Some growers do not give any attention to placing the mushrooms in the baskets. The stems are cut off in the packing room, they are thrown into the weighing pan, and when the beam tips at three, or four, or five pounds, as the case may be, the mushrooms are emptied into the baskets, leveled down, and the baskets closed for shipment. Others use more care in the packing of the mushrooms; especially is this the case on the part of those who pick the mushrooms when they are somewhat larger and more open, though the practice of placing the mushrooms in a basket is followed even by those who pick before the mushrooms are open. In placing them, one mushroom is taken at a time and put stem downward into the basket, until the bottom is covered with one layer, and then successive layers are placed on top of these. The upper layers in the basket then present a very neat and attractive appearance. In thus placing the mushrooms in [Pg 272] the basket, if there are any mushrooms which are quite large, they are placed in the bottom. The custom of the operator here is different from that of the grower of apples, or of other fruit, where the larger and finer samples are often placed on top, the smaller ones being covered below. It is a curious fact, however, that this practice of placing the largest mushrooms below in the basket is due to the fact that usually the larger mushrooms are not considered so marketable.

Figure 238.—View in packing room, Akron "tunnel," N. Y. Mushroom Co.; placing mushrooms in basket. Copyright.

There are several reasons why the larger mushrooms are not considered so desirable or marketable as the medium-sized or smaller ones. In the first place, the larger mushrooms, under certain conditions, especially those grown in house culture at a comparatively high temperature, are apt to be very ripe, so that the gills are black from over-ripe spores, and are thus somewhat unsightly. Those grown at a lower temperature, as is the case in some mines, do not blacken so soon, and are therefore apt to be free from this objection. Another objection, however, is on the part of [Pg 273] the restaurant owner where mushrooms are served. In serving the mushrooms broiled on toast, the medium-sized one is more desirable from the standpoint of the restaurant owner, in that two medium-sized ones might be sufficient to serve two persons, while one quite large one, weighing perhaps the same as the two medium ones, would only be sufficient to serve one person at the same price, unless the large mushroom was cut in two. If this were done, however, the customer would object to being served with half a mushroom, and the appearance of a half mushroom served in this way is not attractive.

Resoiling. — Once or twice a week during the harvesting period all loose earth, broken bits of spawn, free buttons, etc., should be cleaned out where the mushrooms have been picked. These places should be filled with soil and packed down by hand. All young mushrooms that "fog off" should be gathered up clean. Some persons follow the practice of growing a second crop on the same bed from which the first crop has been gathered. The bed is resoiled by placing about two inches of soil over the old soil. The bed is then watered, sometimes with lukewarm water to which a small quantity of nitrate of soda has been added. The large growers, however, usually do not grow a second crop in this way, but endeavor to exhaust the material in the bed by continuous growth.

Use of manure from beds which have failed. — Manure in which the spawn has failed to run is sometimes removed from the bed and mixed with fresh manure, the latter restoring the heat. If the manure was too wet, the moisture content can now be lessened by the use of dry soil.

Cleaning house to prepare for successive crops. — When the crop is harvested, all the material is cleaned out to prepare the beds for the next crop. The material is taken out "clean," and the floors, beds, walls, etc., swept off very clean. In addition, some growers whitewash the floors and all wood-work. Some whitewash only the floors, depending on sweeping the beds and walls very clean. Still others whitewash the floors and wash the walls with some material to kill out the vermin. Some trap or poison the cockroaches, wood-lice, etc., when they appear. Some growers who succeed well for several years, and then fail, believe that the house "gets tired," as they express it, and that the place must rest for a few years before mushrooms can be grown there again. Others grow mushrooms successfully year after year, but employ the best sanitary methods.

Number of crops during a year. — In caves or mines, where the temperature is low, the beds are in process of formation and cropping [Pg 274] continuously. So soon as a bed has been exhausted the material is cleaned out, and new beds are made as fast as the fresh manure is obtained. In houses where the mushrooms cannot be

419

grown during the summer, the crops are grown at quite regular periods, the first crop during fall and early winter, and the second crop during spring. Some obtain the manure and ferment it during August and September, spawning the beds in September and October. Others begin work on the fermentation of the manure in June or July, make up the beds in July and August, spawn, and begin to draw off the crop somewhat earlier. The second crop is prepared for whenever the first one is drawn off, and this varies even in the experience of the same grower, since the rate of the running of the spawn varies from time to time. Sometimes the crop begins to come four or five weeks from the time of planting the spawn. At other times it may be two or three months before the spawn has run sufficiently for the crop to appear. Usually the crop begins to come on well in six to eight weeks. The crop usually lasts for six weeks to two months, or longer.

Productivity of the beds.—One pound of mushrooms from every two square feet of surface is considered a very good crop. Sometimes it exceeds this, the beds bearing one pound for every square foot, though such a heavy yield is rare. Oftener the yield is less than half a pound for a square foot of surface.

Causes of failure.—The beginner should study very carefully the conditions under which he grows his crops, and if failure results, he should attempt to analyze the results in the light of the directions given for the curing of the manure, its moisture content, "sweetness," character of the spawn, temperature, ventilation, etc. While there should be good ventilation, there should not be drafts of air. A beginner may succeed the first time, the second or third, and then may fail, and not know the cause of the failure. But given a good spawn, the right moisture content of the material at time of planting and running of the spawn, the sweet condition, or proper condition of the curing of the manure, proper sanitary conditions, there should be no failure. These are the most important conditions in mushroom culture. After the spawn has run and the crop has begun to come, the beds have been known to freeze up during the winter, and in the spring begin and continue to bear a good crop. After the spawn has run well, beds have accidentally been flooded with water so that manure water would run out below, and yet come on and bear as good a crop as adjoining beds.

Volunteer mushrooms in greenhouses. — Volunteer mushrooms sometimes [Pg 275] appear in greenhouses in considerable quantity. These start from natural spawn in the manure used, or sometimes from the spawn remaining in "spent" mushroom beds which is mixed with the soil in making lettuce beds, etc., under glass. One of the market gardeners at Ithaca used old spawn in this way, and had volunteer mushrooms among lettuce for several years. In making the lettuce beds in the autumn, a layer of fresh horse manure six inches deep is placed in the bottom, and on this is placed the soil mixed with the old, spent mushroom beds. The following year the soil and the manure at the bottom, which is now rotten, is mixed up, and a fresh layer of manure is placed below. In this way the lettuce bed is self-spawned from year to year. About every six years the soil in the bed is entirely changed. This gardener, during the winter of 1900–1, sold $30.00 to $40.00 worth of volunteer mushrooms. Another gardener, in a previous year, sold over $50.00 worth.

Planting mushrooms with other vegetables. — In some cases gardeners follow the practice of inserting a forkful of manure here and there in the soil where other vegetables are grown under glass, and planting in it a bit of spawn.

Mushroom and vegetable house combined. — Some combine a mushroom house and house for vegetables in one, there being a deep pit where several tiers of beds for mushrooms can be built up, and above this the glass house where lettuce, etc., is grown, all at a temperature of about 60° F.

THREE METHODS SUGGESTED FOR GROWING MUSHROOMS IN CELLARS AND SHEDS.

First Method. — Obtain fresh stable horse manure mixed with straw used in bedding the animals. Shake it out, separating the coarse material from the droppings. Put the droppings in a pile two to three feet deep. Pack down firmly. When the heat rises to near 130° F., turn and shake it out, making a new pile. Make the new pile by layers of manure and loam soil, or rotted sod, one part of soil to eight or nine parts of manure. Turn again when the heat rises to near 130° F., and add the same amount of soil. When the temperature is about 100° F., the material is ready for the beds.

Preparing the beds. — Make the beds as described under the paragraph on pages 250–253, or use boxes. Place the coarse litter in the bottom three to four inches deep. On this place three to four inches of the cured material, pack it down, and continue adding material until the bed is ten to fifteen inches deep. Allow the beds to stand, covering them with straw or excelsior if the air in the cellar or shed is such as to dry out the surface.

Test the moisture content according to directions on page 255.

Watch the temperature. Do not let it rise above 130° F. When it is down [Pg 276] to 90° F. or 70° F., if the manure has a "sweetish" or "mushroomy" smell it is ready to spawn.

Spawn according to directions on page 263.

Soil according to directions on page 266; cover bed with straw or excelsior.

Second Method. — Use horse droppings freed from the coarser material. Proceed as in *first* method.

Third Method. — Use horse droppings freed from coarser material. Pile and *pack firmly*. Do not let temperature rise above 130° F. When it has cooled to 100° F., make up the beds, at the same time mixing in an *equal quantity* of rich loam or rotted sod. Spawn in a day or two.

In beginning, practice on a small scale and study the conditions thoroughly, as well as the directions given in this chapter.

[Pg 277]

FOOTNOTES:

[D] I have not learned the history of the other kinds of spawn referred to above.

CHAPTER XXI.

RECIPES FOR COOKING MUSHROOMS.

By MRS. SARAH TYSON RORER.

As varieties of mushrooms differ in analysis, texture and density of flesh, different methods of cooking give best results. For instance, the *Coprinus micaceus*, being very delicate, is easily destroyed by over-cooking; a dry, quick pan of the "mushroom bells" retains the best flavor; while the more dense *Agaricus campestris* requires long, slow cooking to bring out the flavor, and to be tender and digestible. Simplicity of seasoning, however, must be observed, or the mushroom flavor will be destroyed. If the mushroom itself has an objectionable flavor, better let it alone than to add mustard or lemon juice to overcome it. Mushrooms, like many of the more succulent vegetables, are largely water, and readily part with their juices on application of salt or heat; hence it becomes necessary to put the mushroom over the fire usually without the addition of water, or the juices will be so diluted that they will lack flavor. They have much better flavor cooked without peeling, with the exception of puff-balls, which should always be pared. As they lose their flavor by soaking, wash them quickly, a few at a time; take the mushroom in the left hand and with the right hand wash the top or pileus, using either a very soft brush or a piece of flannel; shake them well and put them into a colander to dry.

AGARICUS. [E]

The wild or uncultivated *Agaricus campestris*, which is usually picked in open fields, will cook in less time than those grown in caves and sold in our markets during the winter and spring. Cut the stems close to the gills; these may be put aside and used for flavoring sauces or soups. Wash the mushrooms carefully, keeping the gills down; throw them into a colander until drained.

Stewed.—To each pound, allow two ounces of butter. Put the butter into a saucepan, and when melted, not brown, throw in the mushrooms either whole or cut into slices; sprinkle over a teaspoonful of salt; cover the saucepan closely to keep in the flavor, and [Pg 278] cook very slowly for twenty minutes, or until they are tender. Moisten a rounding tablespoonful of flour in a little cold milk; when perfectly smooth, add sufficient milk to make one gill; stir this into the mushrooms, add a saltspoon of white pepper, stir carefully until boiling, and serve at once. This makes a fairly thick sauce. Less flour is required when they are to be served as a sauce over chicken, steak, or made dishes.

Broiled. — Cut the stems close to the gills; wash the mushrooms and dry them with a soft piece of cheesecloth; put them on the broiler gills up. Put a piece of butter, the size of a marrowfat pea, in the center of each; sprinkle lightly with salt and pepper. Put the broiler over the fire skin side down; in this way, the butter will melt and sort of baste the mushrooms. Have ready squares of neatly toasted bread; and, as soon as the mushrooms are hot on the skin side, turn them quickly and broil about two minutes on the gill side. Five minutes will be sufficient for the entire cooking. Dish on toast and serve at once.

Panned on Cream Toast. — Cut the stem close to the gills; wash and dry as directed for broiling. Put them into a pan, and pour over a very little melted butter, having gill sides up; dust with salt and pepper, run into a hot oven for twenty minutes. While these are panning, toast sufficient bread to hold them nicely; put it onto a hot platter, and just as the mushrooms are done, cover the bread with hot milk, being careful not to have too much or the bread will be pasty and soft. Dish the mushrooms on the toast, putting the skin side up, pour over the juices from the pan, and serve at once.

These are exceedingly good served on buttered toast without the milk, and will always take the place of broiled mushrooms.

In the Chafing Dish. — Wash, dry the mushrooms, and cut them into slices. To each pound allow two ounces of butter. Put the butter in the chafing dish, when hot put in the mushrooms, sprinkle over a teaspoonful of salt, cover the dish, and cook slowly for five minutes, stirring the mushrooms frequently; then add one gill of milk. Cover the dish again, cook for three minutes longer, add the beaten yolks of two eggs, a dash of pepper, and serve at once. These must not be boiled after the eggs are added; but the yolk of egg is by far the most convenient form of thickening when mushrooms are cooked in the chafing dish.

Under the Glass Cover or "Bell" with Cream. — With a small biscuit cutter, cut rounds from slices of bread; they should be about two and a half inches in diameter, and about a half inch in thickness. Cut the stems close to the gills from fresh mushrooms; wash and [Pg 279] wipe the mushrooms. Put a tablespoonful of butter in a saucepan; when hot, throw in the mushrooms, skin side down; cook just a moment, and sprinkle them with salt and pepper. Arrange the rounds of bread, which have been slightly toasted, in the bottom of your "bell" dish; heap the mushrooms on these; put a little piece of butter in the center; cover over the bell, which is either of glass, china, or silver; stand them in a baking pan, and then in the oven for twenty minutes. While these are cooking, mix a tablespoonful of butter and one of flour in a saucepan, add a half pint of milk, or you may add a gill of milk and a gill of chicken stock; stir until boiling, add a half teaspoonful of salt and a dash of pepper. When the mushrooms have been in the oven the allotted time, bring them out; lift the cover, pour over quickly a little of this sauce, cover again, and send them at once to the table.

Another Method. — Wash and dry the mushrooms; arrange them at once on the "bell plate." The usual plates will hold six good sized ones. Dust with pepper and salt; put in the center of the pile a teaspoonful of butter; pour over six tablespoonfuls of cream or milk; cover with the bell; stand the dish in a baking pan, and then in a hot oven for twenty minutes.

These are arranged for individual bells. Where one large bell is used, the mushrooms must be dished on toast before they are served. The object in covering with the bell is to retain every particle of the flavor. The bell is then lifted at the table, that the eater may get full aroma and flavor from the mushroom.

Puree. — Wash carefully a half pound of mushrooms; chop them fine, put them into a saucepan with a tablespoonful of butter, and if you have it, a cup of chicken stock; if not, a cup of water. Cover the vessel and cook slowly for thirty minutes. In a double boiler, put one pint of milk. Rub together one tablespoonful of butter and two tablespoonfuls of flour; add it to the milk; stir and cook until thick; add the mushrooms, and press the whole through a sieve; season to taste with salt and pepper only.

Cream of Mushroom Soup. — This will be made precisely the same as in the preceding recipe, save that one quart of milk will be used instead of a pint with the same amount of thickening, and the mushrooms will not be pressed through a sieve. [Pg 280]

COPRINUS COMATUS and COPRINUS ATRAMENTARIUS.

As these varieties usually grow together and are sort of companion mushrooms, recipes given for one will answer for the cooking of the other. Being soft and juicy, they must be handled with care, and are much better cooked with dry heat. Remove the stems, and wash them carefully; throw them into a colander until dry; arrange them in a baking pan; dot here and there with bits of butter, allowing a tablespoonful to each half pound of mushrooms; dust with salt and pepper, run them into a very hot oven, and bake for thirty minutes; dish in a heated vegetable dish, pouring over the sauce from the pan.

The *C. micaceus* may also be cooked after the same fashion—after dishing the mushrooms boil down the liquor.

Stewed.—Wash and dry them; put them into a large, flat pan, allowing a tablespoonful of butter to each half pound of mushrooms; sprinkle at once with salt and pepper; cover the pan, and stew for fifteen minutes. Moisten a tablespoonful of flour in a little cold milk; when smooth, add a half cup of cream, if you have it; if not, a half cup of milk. Push the mushrooms to one side; turn in this mixture, and stir until boiling. Do not stir the mushrooms or they will fall apart and become unsightly. Dish them; pour over the sauce, and serve at once. Or they may be served on toast, the dish garnished with triangular pieces of toast.

COPRINUS MICACEUS.

Wash and dry the mushrooms; put them into a deep saucepan with a tablespoonful of butter to each quart; stand over a quick fire, sort of tossing the saucepan. Do not stir, or you will break the mushrooms. As soon as they have reached the boiling point, push them to the back part of the stove for five minutes; serve on toast. These will be exceedingly dark, are very palatable, and perhaps are the most easily digested of all the varieties.

LEPIOTA.

These mushrooms, having very thin flesh and deep gills, must be quickly cooked to be good. Remove the stem, take the mushrooms in your hand, gill side down, and with a soft rag wash carefully the top, removing all the little brown scales. Put them into a baking pan, or on a broiler. Melt a little butter, allow it to settle, take the [Pg 281] clear, oily part from the top and baste lightly the mushrooms, gill sides up; dust with salt and pepper. Place the serving dish to heat. Put the mushrooms over a quick fire, skin side down, for just a moment; then turn and boil an instant on the gill side, and serve at once on the heated plate.

In this way *Lepiota procera* is most delicious of all mushrooms; but if cooked in moist heat, it becomes soft, but tough and unpalatable; if baked too long, it becomes dry and leathery. It must be cooked quickly and eaten at once. All the edible forms may be cooked after this recipe.

These are perhaps the best of all mushrooms for drying. In this condition they are easily kept, and add so much to an ordinary meat sauce.

OYSTER MUSHROOMS (Pleurotus).

Wash and dry the mushrooms; cut them into strips crosswise of the gills, trimming off all the woody portion near the stem side. Throw the mushrooms into a saucepan, allowing a tablespoonful of butter to each pint; sprinkle over a half teaspoonful of salt; cover, and cook slowly for twenty minutes. Moisten a tablespoonful of flour in a half cup of milk; when perfectly smooth, add another half cup; turn this into the mushroom mixture; bring to boiling point, add just a grating of nutmeg, a few drops of onion juice, and a dash of pepper. Serve as you would stewed oysters.

To make this into à la poulette, add the yolks of two eggs just as you take the mixture from the fire, and serve on toast.

Mock Oysters. — Trim the soft gill portion of the *Pleurotus ostrea-tus* into the shape of an oyster; dust with salt and pepper; dip in beaten egg, then in bread crumbs, and fry in smoking hot fat as you would an oyster, and serve at once. This is, perhaps, the best method of cooking this variety.

RUSSULA.

While in this group we have a number of varieties, they may all be cooked after one recipe. The stems will be removed, the mushrooms carefully washed, always holding the gill side down in the water, drained in a colander; and while they apparently do not contain less water than other mushrooms, the flesh is rather dense, and they do not so quickly melt upon being exposed to heat. They are nice broiled or baked, or may be chopped fine and served with mayonnaise dressing, stuffed into peeled tomatoes, or with mayonnaise [Pg 282] dressing on lettuce leaves, or mixed with cress and served with French dressing, as salads.

The "green" or *Russula virescens* may be peeled, cut into thin slices, mixed with the leaves of water-cress which have been picked carefully from the stems, covered with French dressing, and served on slices of tomato. It is well to peel all mushrooms if they are to be served raw. To bake, follow recipes given for baking *campestris*. In this way they are exceedingly nice over the ordinary broiled steak.

One of the nicest ways, however, of preparing them for steak is to wash, dry and put them, gills up, in a baking pan, having a goodly quantity; pour over just a little melted butter; dust with salt and pepper, and put them into the oven for fifteen minutes. While you are broiling the steak, put the plate upon which it is to be served over hot water to heat; put on it a tablespoonful of butter, a little salt, pepper, and some finely chopped parsley. Take the mushrooms from the oven, put some in the bottom of the plate, dish the steak on top, covering the remaining quantity over the steak. Add two tablespoonfuls of stock or water to the pan in which they were baked; allow this to boil, scraping all the material from the pan; baste this over the steak, and serve at once.

Agaricus campestris and many other varieties may also be used in this same way.

LACTARII.

Remove the stems, and wash the mushrooms. Put them into a saucepan, allowing a tablespoonful of butter and a half teaspoonful of salt to each pint. Add four tablespoonfuls of stock to the given quantity; cover the saucepan, and *cook slowly* three-quarters of an hour. At the end of this time you will have a rich, brown sauce to which you may add a teaspoonful of Worcestershire sauce, and, if you like, a tablespoonful of sherry. Serve in a vegetable dish.

Lactarius deliciosus Stewed.—Wash the mushrooms; cut them into slices; put them into a saucepan, allowing a half pint of stock to each pint of mushrooms; add a half teaspoonful of salt; cover and stew slowly for three-quarters of an hour. Put a tablespoonful of butter in another saucepan, mix with it a tablespoonful of flour; add the mushrooms, stir until they have reached the boiling point; add a teaspoonful of kitchen bouquet, a dash of pepper, and serve it at once in a heated vegetable dish.

A nice combination for a steak sauce is made by using a dozen good sized *Lactarius deliciosus* with four "beefsteak" mushrooms, using then the first recipe. [Pg 283]

BEEFSTEAK SMOTHERED WITH MUSHROOMS.

Wash a dozen good sized mushrooms, either *Lactarii* or *Agarici*, also wash and remove the spores from half a dozen good sized "beefsteak" mushrooms, cutting them into slices. Put all these into a baking pan, sprinkle over a half teaspoonful of salt, add a tablespoonful of butter, and bake in a moderate oven three-quarters of an hour. Broil the steak until it is nearly done; then put it into the pan with the mushrooms, allowing some of the mushrooms to remain under the steak, and cover with the remaining portion; return it to the oven for ten minutes; dish and serve at once.

BOLETI.

These are more palatable baked or fried. Wash the caps and remove the pores. Dip the caps in beaten egg, then in bread crumbs, and fry them in smoking hot fat; oil is preferable to butter; even suet would make a drier fry than butter or lard. Serve at once as you would egg plant.

Baked. — Wash and remove the pores; put the mushrooms into a baking pan; baste them with melted butler, dust with salt and pepper, and bake in a moderately hot oven three-quarters of an hour; dish in a vegetable dish. Put into the pan in which they were baked, a tablespoonful of butter. Mix carefully with a tablespoonful of flour and add a half pint of stock, a half teaspoonful of kitchen bouquet or browning, the same of salt, and a dash of pepper; pour this over the mushrooms, and serve.

In Fritter Batter. — Beat the yolk of one egg slightly, and add a half cup of milk; stir into this two-thirds of a cup of flour; stir in the well beaten white of the egg and a teaspoonful of olive oil. Wash and remove the pores from the boleti. Have ready a good sized shallow pan, the bottom covered with smoking hot oil; dip the mushrooms, one at a time, into this batter, drain for a moment, and drop them into the hot fat. When brown on one side, turn and brown on the other. Drain on soft paper and serve at once.

Boleti in Brown Sauce. — Wash and dry the boleti; remove the pores; cut them into small pieces. To each pound allow a tablespoonful of butter. Put the butter into a saucepan with the mushrooms; add a half teaspoonful of salt; cover the pan, and stew slowly for twenty minutes; then dust over a tablespoonful of flour; add a half cup of good beef stock; cook slowly for ten minutes longer, and serve. [Pg 284]

HYDNUM.

As these mushrooms are slightly bitter, they must be washed, dried, and thrown into a little boiling water, to boil for just a moment; drain, and throw away this water, add a tablespoonful of butter, a teaspoonful of salt, a dash of pepper, and a half cup of milk or stock; cover the pan, and cook slowly for twenty minutes. As the milk scorches easily, cook over a very slow fire, or in a double boiler. Pour the mixture over slices of toast, and serve at once. A tablespoonful or two of sherry may be added just as they are removed from the fire.

CLAVARIA.

Wash, separating the bunches, and chop or cut them rather fine, measure, and to each quart allow a half pint of Supreme sauce. Throw the clavaria into a saucepan, cover, and allow it to stew gently for fifteen minutes while you make the sauce. Put a tablespoonful of butter and one of flour in the saucepan; mix, and add a half pint of milk or chicken stock; or you may add half of one and half of the other; stir until boiling; take from the fire, add a half teaspoonful of salt, a saltspoonful of pepper, and the yolks of two eggs. Take the clavaria from the fire, and when cool stir it into the sauce. Turn into a baking dish, sprinkle the top with crumbs, and brown in a quick oven. Do not cook too long, as it will become watery.

Pickled Clavaria.—Wash the clavaria thoroughly without breaking it apart; put into a steamer; stand the steamer over a kettle of boiling water, and steam rapidly, that is, keep the water boiling hard for fifteen minutes. Take from the fire, and cool. Put over the fire sufficient vinegar to cover the given quantity; to each quart, allow two bay leaves, six cloves, a teaspoonful of whole mustard, and a dozen pepper corns, that is, whole peppers. Put the clavaria into glass jars. Bring the vinegar to boiling point, and pour it over; seal and put aside.

This may be served alone as any other pickle, or on lettuce leaves with French dressing as a salad.

Escalloped Clavaria.—Wash, separate and cut the clavaria as in first recipe. To each quart allow a half pint of chicken stock, a teaspoonful of salt, a tablespoonful of chopped parsley. Put a layer of bread crumbs in the bottom of the dish, then a layer of chopped clavaria, and so continue until you have the dish filled. Pour over the stock, which you have seasoned with salt and pepper; dot bits of butter [Pg 285] here and there over the top, and bake in a moderate oven thirty minutes.

This recipe is excellent for the young or button *Hypholoma*, except that the time of baking must be forty-five minutes.

PUFF-BALLS.

To be eatable, the puff-balls must be perfectly white to the very center. Pare off the skin; cut them into slices; dust with salt and pepper. Have ready in a large, shallow pan a sufficient quantity of hot oil to cover the bottom. Throw in the slices and, when brown on one side, turn and brown on the other; serve at once on a heated dish.

A la Poulette.—Pare the puff-balls; cut them into slices and then into dice; put them into a saucepan, allowing a tablespoonful of butter to each pint of blocks. Cover the saucepan; stew gently for fifteen minutes; lift the lid; sprinkle over a teaspoonful of salt and a dash of pepper. Beat the yolks of three eggs until light; add a half cup of cream and a half cup of milk; pour this into the hot mixture, and shake until smoking hot. Do not allow them to boil. Serve in a heated vegetable dish, with blocks of toast over the top.

Puff-Ball Omelet.—Pare and cut into blocks sufficient puff-balls to make a pint. Put a tablespoonful of butter into a saucepan; add the puff-balls, cover and cook for ten minutes. Beat six eggs without separating, until thoroughly mixed, but not too light; add the cooked puff-balls, a level teaspoonful of salt and a dash of pepper. Put a tablespoonful of butter into your omelet pan; when hot, turn in the egg mixture; shake over the hot fire until the bottom has thoroughly set, then with a limber knife lift the edge, allowing the soft portion to run underneath; continue this operation until the omelet is cooked through; fold and turn onto a heated dish. Serve at once.

Other delicate mushrooms may be used in this same manner.

Puff-Balls with Agaricus campestris.—As the *Agaricus campestris* has a rather strong flavor and the puff-balls are mild, both are better for being mixed in the cooking. Take equal quantities of *Agaricus campestris* and puff-balls; pare and cut the puff-balls into blocks; to each half pound allow a tablespoonful of butter. Put the butter in a saucepan, add the mushrooms, sprinkle over the salt (allowing a half teaspoonful always to each pint); cover the saucepan and stew slowly for twenty minutes. Moisten a tablespoonful of flour in a half cup of milk, add it to the mixture, stir and cook for just a moment, add a dash of pepper, and serve in a heated dish. [Pg 286]

This recipe may be changed by omitting the flour and adding the yolks of a couple of eggs; milk is preferable to stock, for all the white or light-colored varieties.

MORCHELLA.

Select twelve large-sized morels; cut off the stalks, and throw them into a saucepan of warm water; let them stand for fifteen minutes; then take them on a skimmer one by one, and drain carefully. Chop fine sufficient cold boiled tongue or chicken to make one cupful; mix this with an equal quantity of bread crumbs, and season with just a suspicion of onion juice, not more than ten drops, and a dash of pepper. Fill this into the mushrooms, arrange them neatly in a baking pan, put in a half cup of stock and a tablespoonful of butter, bake in a moderate oven thirty minutes, basting frequently. When done, dish neatly. Boil down the sauce that is in the pan until it is just sufficient to baste them on the dish; serve at once.

A Second Method. — Select large-sized morels; cut off the stalk; wash well through several waters. Put into a frying pan a little butter, allowing about a tablespoonful to each dozen mushrooms. When hot, throw in the mushrooms, and toss until they are thoroughly cooked; then add a half pint of milk or stock; cover the vessel, and cook slowly twenty minutes; dust with salt and pepper, and serve in a vegetable dish. This method gives an exceedingly palatable and very sightly dish if garnished with sweet Spanish peppers that have been boiled until tender.

Another Method. — Remove the stems, and wash the morels as directed in the preceding recipe. Make a stuffing of bread crumbs seasoned with salt, pepper, chopped parsley, and sufficient melted butter to just moisten. Place them in a baking pan; add a little stock and butter; bake for thirty minutes. When done, dish. Into the pan in which they were cooked, turn a cupful of strained tomatoes; boil rapidly for fifteen minutes until slightly thickened; pour this over the mushrooms; garnish the dish with triangular pieces of toasted bread, and serve.

GENERAL RECIPES.

In the following recipes one may use *Agaricus campestris*, *silvicola*, *arvensis*, or *Pleurotus ostreatus*, or *sapidus*, or *Coprinus comatus*, or any kindred mushrooms. The *Agaricus campestris*, however, are to be preferred.

To Serve with a Boiled Leg of Mutton, wash well the mushrooms and [Pg 287] dry them; dip each into flour, being careful not to get too much on the gill side. In a saucepan have a little hot butter or oil; drop these in, skin side down; dust them lightly with salt and pepper. After they have browned on this side, turn them quickly and brown the gills; add a half pint of good stock; let them simmer gently for fifteen minutes. Take them up with a skimmer, and dish them on a platter around the mutton. Boil the sauce down until it is the proper consistency; pour it over, and serve at once. These are also good to serve with roasted beef.

Mushroom Sauce for Game.—Wash well one pound of fresh mushrooms; dry, and chop them very fine. Put them into a saucepan with one and a half tablespoonfuls of butter; cover, and cook slowly for eight minutes; then add a half cup of fresh rubbed bread crumbs, a half teaspoonful of salt, a saltspoon of white pepper; cover and cook again for five minutes; stir, add a tablespoonful of chopped parsley, and, if you like, two tablespoonfuls of sherry; turn into a sauce-boat.

A Nice Way to Serve with Fricassee of Chicken.—Wash and dry the mushrooms; sprinkle them with salt and pepper. Put some oil or butter in a shallow pan; when hot, throw in the mushrooms, skin side down; cover the pan, put in the oven for fifteen minutes; baste them once during the baking. Lift them carefully and put them on a heated dish. Add to the fat in the pan two tablespoonfuls of finely chopped mushrooms, a half cup of good stock; boil carefully for five minutes. Have ready rounds of bread toasted; dish the mushrooms on these; put on top a good sized piece of carefully boiled marrow; season the sauce with salt, and strain it over. Use these as a garnish around the edge of the plate, or you may simply dish and serve them for breakfast, or as second course at lunch.

Oysters and Mushrooms.—Wash and remove the stems from a half pound of fresh mushrooms; chop them fine; put them into a saucepan with a tablespoonful of butter, a half teaspoonful of salt, and a dash of pepper; cover closely, and cook over a slow fire for ten minutes. Have ready, washed and drained, twenty-five good sized fat oysters; throw them perfectly dry into this mushroom mixture. Pull the saucepan over a bright fire; boil, stirring carefully, for about five minutes. Serve on squares of carefully toasted bread.

Tomatoes Stuffed with Mushrooms.—Wash perfectly smooth, solid tomatoes; cut a slice from the stem end, and remove carefully the seeds and core. To each tomato allow three good sized mushrooms; wash, dry, chop them fine, and stuff them into the tomatoes; put a half saltspoon of salt on the top of each and a dusting of pepper. [Pg 288] Into a bowl put one cup of soft bread crumbs; season it with a half teaspoonful of salt and a dash of pepper; pour over a tablespoonful of melted butter; heap this over the top of the tomato, forming a sort of pyramid, packing in the mushrooms; stand the tomatoes in a baking pan and bake in a moderate oven one hour. Serve at once, lifting them carefully to prevent breaking.

Or, the mushrooms may be chopped fine, put with a tablespoonful of butter into a saucepan and cooked for five minutes before they are stuffed into the tomatoes; then the bread crumbs packed over the top, and the whole baked for twenty minutes. Each recipe will give you a different flavor.

FOOTNOTES:

[E] The recipes for Agaricus are intended for the several species of this genus (Psalliota).

CHAPTER XXII.

CHEMISTRY AND TOXICOLOGY OF MUSHROOMS.

By J. F. CLARK.

Regarding the chemical composition of mushrooms, we have in the past been limited largely to the work of European chemists. Recently, however, some very careful analyses of American mushrooms have been made. The results of these investigations, while in general accord with the work already done in Europe, have emphasized the fact that mushrooms are of very variable composition. That different species should vary greatly was of course to be expected, but we now know that different specimens of the same species grown under different conditions may be markedly different in chemical composition. The chief factors causing this variation are the composition, the moisture content, and the temperature of the soil in which they grow, together with the maturity of the plant. The temperature, humidity, and movement of the atmosphere and other local conditions have a further influence on the amount of water present.

The following table, showing the amounts of the more important constituents in a number of edible American species, has been compiled chiefly from a paper by L. B. Mendel (Amer. Jour. Phy. **1**: 225–238). This article is one of the most recent and most valuable contributions to this important study, and anyone wishing to look into the methods of research, or desiring more detailed information than is here given, is referred to the original paper. [Pg 289]

Table I.

FRESH MATERIAL.		IN WATER-FREE MATERIAL.			
WATER.	DRY MATTER.	TOTAL NITROGEN.	PROTEID NITROGEN.	ETHER EXTRACT.	SOLUBLE IN 85 PER CENT ALCOHOL.
%	%	%	%	%	%
92.19	7.81	5.79	1.92	3.3	56.3
89.54	0.46	4.66	3.49	4.8	29.3
70.80	9.20	3.29	2.23	3.2	27.8
73.70	6.30	2.40	1.13	1.6	31.5

FRESH MATERIAL.		IN WATER-FREE MATERIAL.				
WATER.	DRY MATTER.	TOTAL NITROGEN.	PROTEID NITROGEN.	ETHER EXTRACT.	SOLUBLE IN 85 PER CENT AL-COHOL.	FI
89.61	0.39	5.36	1.98	6.0	57.2	
88.97	1.03	4.28	2.49	2.5	44.4	
91.8	8.2	4.75	3.57	3.72	– –	

[Pg 290] **Water.** — Like all growing plants, the mushroom contains a very large proportion of water. The actual amount present varies greatly in different species. In the above table it will be seen that *Polyporus sulphureus*, with over 70 per cent. of water, has the least of any species mentioned, while the species of *Coprinus* and *Agaricus* have usually fully 90 per cent. water. The amount of water present, however, varies greatly in the same species at different seasons and in different localities, and with variations in the moisture content of soil and atmosphere, also with the age and rapidity of development of the individual plant.

Total Nitrogen. — The proportion of nitrogen in the dry matter of different species varies from 2 per cent. to 6 per cent. This comparatively high nitrogen content was formerly taken to indicate an unusual richness in proteid substances, which in turn led to very erroneous ideas regarding the nutritive value of these plants. The nitrogenous substances will be more fully discussed later, when we consider their nutritive value.

Ether Extract. — This consists of a variety of fatty substances soluble in ether. It varies greatly in quality and quantity in different species. The amount is usually from 4 per cent. to 8 per cent. of the total dry matter. It includes, besides various other substances, several free fatty acids and their glycerides, the acids of low melting point being most abundant. These fatty substances occur in the stem, but are much more abundant in the cap, especially in the fruiting portion. Just what nutritive value these fatty matters may have has never been determined.

Carbohydrates. — The largest part of the dry matter of the mushrooms is made up of various carbohydrates, including cellulose or fungocellulose, glycogen, mycoinuline, trehalose, mannite, glucose, and other related substances. The cellulose is present in larger proportion in the stem than in the cap, and in the upper part of the cap than in the fruiting surface. This is doubtless related to the sustaining and protective functions of the stem and the upper part of the cap. Starch, so common as a reserve food in the higher plants, does not occur in the mushrooms. As is the case with the fats, no determination of the nutritive value of these substances has been made, but it may be assumed that the soluble carbohydrates of the mushrooms do not differ greatly from similar compounds in other plants.

Ash. — The ash of mushrooms varies greatly. *Polyporus officinalis* gives but 1.08 per cent. of ash in dry matter, *Pleurotus ulmarius* gives 12.6 per cent., and *Clitopilus prunulus* gives 15 per cent. The average [Pg 291] of twelve edible species gave 7 per cent. ash in the stem and 8.96 per cent. in the cap.

In regard to the constituents of the ash, potassium is by far the most abundant — the oxide averaging about 50 per cent. of the total ash. Phosphoric acid stands next to potassium in abundance and importance, constituting, on an average, about one-third of the entire ash. Oxides of manganese and iron are always present; the former averaging about 3 per cent. and the latter 5 per cent. to 2 per cent. of the ash. Sodium, calcium, and chlorine are usually present in small and varying quantities. Sulphuric acid occurs in the ash of all fungi, and is remarkable for the great variation in quantity present in different species; e. g., ash of *Helvella esculenta* contains 1.58 per cent. H_2SO_4 while that of *Agaricus campestris* contains the relatively enormous amount of 24.29 per cent.

Any discussion of the bare composition of a food is necessarily incomplete without a consideration of the nutritive value of the various constituents. This is especially desirable in the case of the mushrooms, for while they are frequently overestimated and occasionally ridiculously overpraised by their friends, they are quite generally distrusted and sometimes held in veritable abhorrence by those who are ignorant of their many excellent qualities. On the one hand, we are told that "gastronomically and chemically considered the flesh of the mushroom has been proven to be almost identical with meat, and possesses the same nourishing properties." We frequently hear them referred to as "vegetable beefsteak," "manna of the poor," and other equally extravagant and misleading terms. On the other hand, we see vast quantities of the most delicious food rotting in the fields and woods because they are regarded by the vast majority of the people as "toadstools" and as such particularly repulsive and poisonous.

Foods may be divided into three classes according to the functions they perform:

- (*a*) To form the material of the body and repair its wastes.
- (*b*) To supply energy for muscular exertion and for the maintenance of the body heat.
- (*c*) Relishes.

The formation of the body material and the repair of its wastes is the function of the proteids of foods. It has been found by careful experiment that a man at moderately hard muscular exertion requires .28 lb. of digestible proteids daily. The chief sources of our proteid foods are meats, fish, beans, etc. It has been as a proteid food that mushrooms have been most strongly recommended. Referring [Pg 292] to Table I, it will be seen that nitrogen constituted 5.79 per cent. of the total dry substance of *Coprinus comatus*. This high nitrogen content, which is common to the mushrooms in general, was formerly taken to indicate a very unusual richness in proteid materials. It is now known, however, that there were several sources of error in this assumption.

Much of the nitrogen is present in the form of non-proteid substances of a very low food value. Another and very considerable portion enters into the composition of a substance closely related to cellulose. A third source of error was the assumption that all the proteid material was digestible. It is now known that a very considerable portion is not digestible and hence not available as food. Thus, notwithstanding the 5.79 per cent. of nitrogen in *Coprinus comatus*, we find but .82 per cent. in the form of actually available (i. e., digestible) proteids, or approximately one-seventh of what was formerly supposed to be present.

The digestibility of the proteids varies very greatly with the species. Mörner found the common field mushroom, *Agaricus campestris*, to have a larger amount of proteids available than any other species studied by him. Unfortunately, the digestibility of the American plant has not been tested. There is great need for further work along this line. Enough has been done, however, to demonstrate that mushrooms are no longer to be regarded as a food of the proteid class.

The energy for the muscular exertion and heat is most economically derived from the foods in which the carbohydrates and fats predominate.

The common way of comparing foods of the first two classes scientifically is to compare their heat-giving powers. The unit of measurement is termed a *calorie*. It represents the amount of heat required to raise a kilogram of water 1° Centigrade. (This is approximately the heat required to raise one pound of water 4° Fahrenheit.) A man at moderately hard muscular labor requires daily enough food to give about 3500 *calories* of heat-units. The major part of this food may be most economically derived from the foods of the second class, any deficiency in the .28 lb. of digestible protein being made up by the addition of some food rich in this substance.

In the following table the value of ten pounds of several food substances of the three classes has been worked out. Especial attention is called to the column headed "proteids" and to the last column where the number of heat-units which may be purchased for one cent at current market rates has been worked out. [Pg 293]

Table II.

NUTRITIVE VALUE OF TEN POUNDS OF SEVERAL FOODS.

	PROTEIDS.	FATS.	CARBO-HYDRATES.	CALORIES.	COST.	CALORIES FOR ONE CENT.
Beef (round)	1.87	.88	— —	7200	$1.50	48.
Beans (dried)	2.23	.18	5.91	15900	.30	530.
Cabbage	.18	.03	.49	1400	.15	93.
Potatoes	.18	.01	1.53	3250	.10	325.
Flour (roller process)	1.13	.11	7.46	16450	.25	658
Coprinus comatus	.04	.025	.434	987	2.50	3.9
Pleurotus ostreatus	.051	.042	.828	1811	2.50	7.2
Morchella esculenta	.094	.05	.306	955	2.50	3.8
Agaricus campestris	.18	.03	.46	1316	2.50	5.3
Oysters	.61	.14	.33	2350	2.00	11.7

[Pg 294] The mushrooms have been valued at 25 cents per pound, which is probably considerably below the average market price for a good article. It should also be remarked that the amounts given in this table are the digestible and hence available constituents of the foods. The only exception to this is in the case of the fats and carbohydrates of the mushrooms, no digestion experiments having been reported on these constituents. In the absence of data we have assumed that they were entirely digested.

The beef and beans are typical animal and vegetable foods of the proteid class. A glance at the table will show how markedly they differ from the mushrooms. The latter are nearest the cabbage in composition and nutritive value. The similarity between the cabbage and the *Agaricus campestris* here analyzed is very striking. The potato is somewhat poorer in fat, but very much richer than the mushroom in carbohydrates.

The figures in the last column will vary of course with fluctuations in the market price, but such variation will not interfere at any time with the demonstration that *purchased* mushrooms are not a poor man's food. Here we find that one cent invested in cabbage at 1-1/2 cents per pound, gives 93 *calories* of nutrition, while the same amount invested in *Agaricus campestris* — the common mushroom of our markets — would give but 5.3 *calories*, although they are almost identical so far as nutritive value is concerned.

The same sum invested in wheat flour, with its high carbohydrate and good proteid content, would yield 658 *calories* or one-sixth the amount necessary to sustain a man at work for one day. The amount of mushrooms necessary for the same result is a matter of simple computation.

Mushrooms, however, have a distinct and very great value as a food of the third class, that is, as condiments or food accessories, and their value as such is beyond the computation of the chemist or the physiologist, and doubtless varies with different individuals. They are among the most appetizing of table delicacies and add greatly to the palatability of many foods when cooked with them. It is surely as unfair to decry the mushroom on account of its low nutritive value, as it is wrong to attribute to it qualities which are nothing short of absurd in view of its composition. In some respects its place as a food is not unlike that of the oyster, celery, berries, and other delicacies. Worked out on the basis of nutritive value alone they would all be condemned; the oyster for instance presents a showing but little better than the mushroom, and vastly inferior, so far as economy is concerned, to the common potato. This, too, for oysters [Pg 295] purchased by the quart. The nutritive value of one cent's worth of oysters "on the half shell" would be interesting!

The question of the toxicology of the higher fungi is one of very great theoretical and practical interest. But on account of the great difficulties in the way of such investigations comparatively little has yet been accomplished. A few toxic compounds belonging chiefly to the class termed alkaloids have, however, been definitely isolated.

Choline.—This alkaloid is of wide occurrence in the animal and vegetable kingdoms. It has been isolated from *Amanita muscaria, A. pantherina, Boletus luridus,* and *Helvella esculenta.* It is not very toxic, but on uniting with oxygen it passes over to muscarine. According to Kobert the substance formed from choline on the decay of the mushrooms containing it is not muscarine, but a very closely related alkaloid, *neurin.* This transformation of a comparatively harmless alkaloid to an extremely deadly one simply by the partial decay of the plant in which the former is normally found, emphasizes very much the wisdom of rejecting for table use all specimens which are not entirely fresh. This advice applies to all kinds of mushrooms, and to worm-eaten and otherwise injured, as well as decayed ones. Neurin is almost identical in its physiological effects with muscarine, which is described below.

Muscarine.—This is the most important because the most dangerous alkaloid found in the mushrooms. It is most abundant in *Amanita muscaria,* it is also found in considerable quantity in *Amanita pantherina,* and to a lesser, but still very dangerous extent in *Boletus luridus* and *Russula emetica.* It is quite probably identical with bulbosine, isolated from *Amanita phalloides* by Boudier. *Muscarine* is an extremely violent poison, .003 to .005 of a gram (.06 grain) being a very dangerous dose for a man. Like other constituents of mushrooms, the amount of muscarine present varies very greatly with varying conditions of soil and climate. This, indeed, may account for the fact that *Boletus luridus* is regarded as an edible mushroom in certain parts of Europe, the environment being such that little or no muscarine is developed.

According to Kobert, *Amanita muscaria* contains, besides choline and muscarine, a third alkaloid, *pilz-atropin.* This alkaloid, like ordinary atropin, neutralizes to a greater or less extent the muscarine.

The amount of pilz-atropin present varies, as other constituents of mushrooms vary, with varying conditions of soil, climate, etc., and it may be that in those localities where the *Amanita muscaria* is used for food the conditions are favorable for a large production of pilz-atropin which neutralizes the muscarine, thus making [Pg 296] the plant harmless. Be this as it may, *Amanita muscaria*, so deadly as ordinarily found, is undoubtedly used quite largely as food in parts of France and Russia, and it has been eaten repeatedly in certain localities in this country without harm.

Fortunately muscarine has a very unpleasant taste. It is interesting in this connection to note that the *Amanita muscaria* is said to be used by the inhabitants of Northern Russia—particularly the Koraks—as a means of inducing intoxication. To overcome the extremely unpleasant taste of the plant they swallow pieces of the dried cap without chewing them, or boil them in water and drink the decoction with other substances which disguise the taste.

The symptoms of poisoning with muscarine are not at once evident, as is the case with several of the less virulent poisons. They usually appear in from one-half to two hours. For the symptoms in detail we shall quote from Mr. V. K. Chestnut, Dept. of Agr., Washington (Circular No. 13, Div. of Bot.): "Vomiting and diarrhœa almost always occur, with a pronounced flow of saliva, suppression of the urine, and various cerebral phenomena beginning with giddiness, loss of confidence in one's ability to make ordinary movements, and derangements of vision. This is succeeded by stupor, cold sweats, and a very marked weakening of the heart's action. In case of rapid recovery the stupor is short and usually marked with mild delirium. In fatal cases the stupor continues from one to two or three days, and death at last ensues from the gradual weakening and final stoppage of the heart's action."

The treatment for poisoning by muscarine consists primarily in removing the unabsorbed portion of the mushroom from the alimentary canal and in counteracting the effect of muscarine on the heart. The action of this organ should be fortified at once by the subcutaneous injection, by a physician, of atropine in doses of from one one-hundredth to one-fiftieth of a grain. The strongest emetics, such as sulphate of zinc or apomorphine, should be used, though in case of profound stupor even these may not produce the desired action. Freshly ignited charcoal or two grains of a one per cent. alkaline solution of permanganate of potash may then be administered, in order, in the case of the former substance, to absorb the poison, or, in the case of the latter, to decompose it. This should be followed by oils or oleaginous purgatives, and the intestines should be cleaned and washed out with an enema of warm water and turpentine.

Experiments on animals poisoned by *Amanita muscaria* and with pure muscarine show very clearly that when the heart has nearly ceased to beat it may be stimulated to strong action almost instantly [Pg 297] by the use of atropine. Its use as thus demonstrated has been the means of saving numerous lives. We have in this alkaloid an almost perfect physiological antidote for muscarine, and therefore in such cases of poisoning its use should be pushed as heroically as the symptoms of the case will warrant. The presence of phallin in *Amanita muscaria* is possible, and its symptoms should be looked for in the red color of the blood serum discharged from the intestines.

Phallin. — The exact chemical nature of this extremely toxic substance is not certainly known, but it is generally conceded to be of an albuminous nature. That it is an extremely deadly poison is shown by the fact that .0015 grain per 2 lbs. weight of the animal is a fatal dose for cats and dogs. It is the active principle of the most deadly of all mushrooms, the *Amanita phalloides*, or death-cup fungus. We quote again from Mr. Chestnut's account of phallin and its treatment: "The fundamental injury is not due, as in the case of muscarine, to a paralysis of the nerves controlling the action of the heart, but to a direct effect on the blood corpuscles. These are quickly dissolved by phallin, the blood serum escaping from the blood vessels into the alimentary canal, and the whole system being rapidly drained of its vitality. No bad taste warns the victim, nor do the preliminary symptoms begin until nine to fourteen hours after the poisonous mushrooms are eaten. There is then considerable abdominal pain and there may be cramps in the legs and other nervous phenomena, such as convulsions, and even lockjaw or other kinds of tetanic spasms. The pulse is weak, the abdominal pain is rapidly followed by nausea, vomiting, and extreme diarrhœa, the intestinal discharges assuming the 'rice-water' condition characteristic of cholera. The latter symptoms are persistently maintained, generally without loss of consciousness, until death ensues, which happens in from two to four days. There is no known antidote by which the effects of phallin can be counteracted. The undigested material, if not already vomited, should, however, be removed from the stomach and intestines by methods similar to those given for cases of poisoning by *Amanita muscaria.*

"After that the remainder of the poison, if the amount of phallin already taken up by the system is not too large, may wear itself out on the blood and the patient may recover. It is suggested that this wearing-out process may be assisted by transfusing into the veins blood freshly taken from some warm-blooded animal. The depletion of the blood serum might be remedied by similar transfusions of salt and warm water."

Helvellic Acid.—This very deadly poison is sometimes found in [Pg 298] *Helvella esculenta* Persoon (Gyromitra esculenta), particularly in old or decaying specimens. It has been studied and named by Boehm. It is quite soluble in hot water, and in some localities this species of *Helvella* is always parboiled—the water being thrown away—before it is prepared for the table. It seems to be quite generally agreed that young and perfectly fresh specimens are free from the poison. As the poison is very violent, however, this plant should be carefully avoided.

The symptoms resemble in a very marked degree those of the deadly phallin, the dissolution of the red corpuscles of the blood being one of the most marked and most dangerous; this is accompanied by nausea, vomiting, jaundice, and stoppage of the kidneys. There is no known antidote for this poison, hence the little that can be done would be similar to that mentioned under phallin.

When poisoning by mushrooms is suspected, one cannot too strongly urge that the services of a competent physician should be secured with the least possible delay.

CHAPTER XXIII.

DESCRIPTION OF TERMS APPLIED TO CERTAIN STRUCTURAL CHARACTERS OF MUSHROOMS.

By H. HASSELBRING.

In fungi, as in higher plants, each organ or part of the plant is subject to a great number of variations which appeal to the eye of the student, and by which he recognizes relationship among the various individuals, species, and genera of this group. For the purpose of systematic studies of mushrooms or even for the recognition of a few species, it is of primary importance to be acquainted with terms used in describing different types of variation. Only a few of the more important terms, such as are employed in this book, together with diagrams illustrating typical cases to which they are applied, will be given here.

The pileus. — The *pileus* or *cap* is the first part of a mushroom which attracts the attention of the collector. It is the fleshy fruit body of the plant. This, like all other parts of the mushroom, is made up, not of cellular tissue as we find it in flowering plants, but of numerous interwoven threads, called *hyphæ*, which constitute the flesh or *trama* of the pileus. Ordinarily, the filamentous structure of the [Pg 299] flesh is very obvious when a thin section of the cap is examined under the microscope, but in certain genera, as *Russula* and *Lactarius*, many branches of the *hyphæ* become greatly enlarged, forming little vesicles or bladders. These vesicles lie in groups all through the flesh of the pileus, sometimes forming the greater part of its substance. The filamentous *hyphæ* pass around and through these groups, filling up the interstices. In cross section this tissue resembles parenchyma, and appears as if it were made up of rounded cells. Such a trama is said to be *vesiculose* to distinguish it from the ordinary or *floccose* trama. The threads on the outer surface of the pileus constitute the cortex or cuticle. They are thick walled and often contain coloring matter which gives the plants their characteristic color. In many species their walls become gelatinized, covering the outside of the pileus with a viscid, slimy, or glutinous layer, often called *pellicle*. In other instances the corticle layer ceases to grow with the pileus. It is then torn and split by the continued expanding of the rest of the plant, and remains on the surface in the form of hairs, fibers, scales, etc.

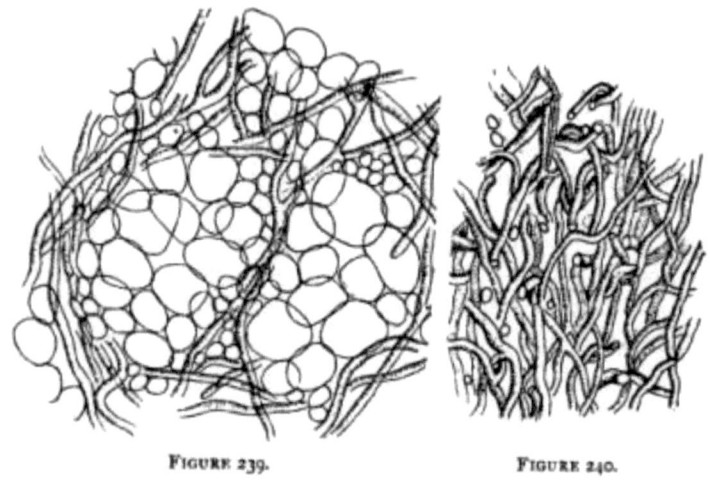

Figure 239.   Figure 240.

Portion of vesiculose trama in the pileus of a Russula.

Portion of a floccose trama.

As an example of the most usual form of the pileus, we may take that of the common mushroom (*Agaricus campestris*) when it is nearly expanded. The pileus is then quite regular in outline and evenly *convex* (Fig. 243). Many mushrooms during the early stages of their development have this form, which is variously changed by later [Pg 300] growth. The convex pileus usually becomes *plane* or *expanded* as it grows. If the convexity is greater it is said to be *campanulate* (Fig. 245), *conical hemispherical*, etc., terms which need no explanation. The pileus is *umbilicate* when it has an abrupt, sharp depression at the center (Fig. 241), *infundibuliform* when the margin is much higher than the center, so that the cap resembles a funnel (Fig. 244), and *depressed* when the center is less, or irregularly, sunken. When the center of the pileus is raised in the form of a boss or knob it is *umbonate* (Fig. 242). The umbo may have the form of a sharp elevation at the center, or it may be rounded or obtuse, occupying a larger part of the disc. When it is irregular or indistinct the pileus is said to be *gibbous* (Fig. 246).

FIGURE 241. FIGURE 242. FIGURE 243.

- Figure 241.—Omphalia campanella, pileus umbilicate, gills decurrent.
- Figure 242.—Lepiota procera, pileus convex, umbonate; annulus free, movable; gills free.
- Figure 243.—Agaricus campestris, pileus convex, gills free.

The gills.—The *gills* or *lamellæ* are thin blades on the under side [Pg 301] of the pileus, radiating from the stem to the margin. When the pileus is cut in halves the general outline of the gills may be observed. In outline they may be broad, narrow, lanceolate, triangular, etc. In respect to their ends they are *attenuate* when gradually narrowed to a sharp point, *acute* when they end in a sharp angle, and *obtuse* when the ends are rounded. Again, the gills are *arcuate*

449

when they arch from the stem to the edge of the pileus, and *ventricose* when they are bellied out vertically toward the earth.

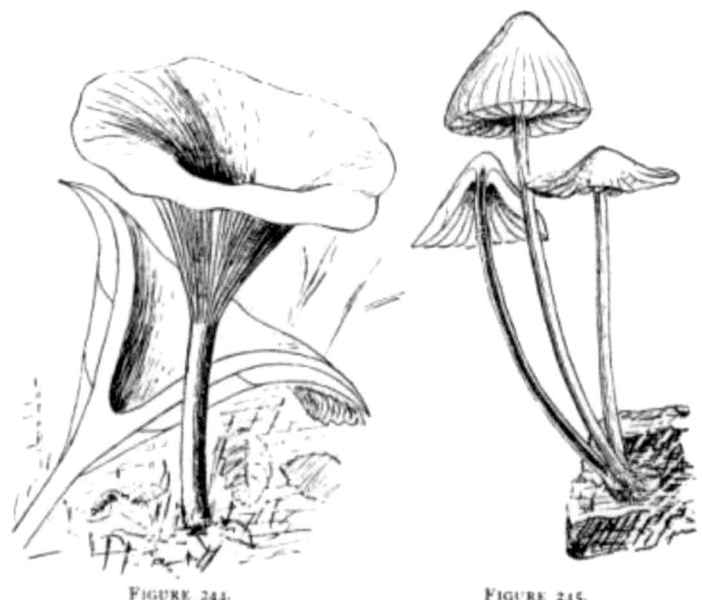

FIGURE 244.

Figure 244.

FIGURE 245.

Figure 245.

Clitocybe infundibuliformis, pileus infundibuliform, gills decurrent.

Mycena galericulata, pileus conic to campanulate, gills decurrent by a tooth, stem fistulose.

The terms given above are often used in descriptive works, but the most important feature to be noted in the section of the plant is the relation of the gills to the stem. This relation is represented by several distinct types which are sometimes used to limit genera or sub-genera, since the mode of attachment is usually constant in all species of a group. The principal relations of the gills to the stem are described as follows: *Adnate* when they reach the stem and are set squarely against it (Fig. 247); *decurrent* when they run down the stem (Fig. 244); *sinuate* or *emarginate* when they have a notch or vertical curve at the posterior end (Fig. 246); and *free* when they [Pg 302] are rounded off without reaching the stem (Fig. 243). In all cases when the lamellæ reach the stem and are only attached by the upper angle they are said to be *adnexed*. This term is often used in combination with others, as *sinuate-adnexed* (Fig. 248, small figure), or *ascending adnexed* (Fig. 248, larger plant). Sometimes the lamellæ are adnate, adnexed, etc., and have a slight decurrent process or tooth as in *Mycena galericulata* (Fig. 245). In many plants the gills separate very readily from the stem when the plants are handled. Sometimes merely the expansion of the pileus tears them away, so that it is necessary to use great caution, and often to examine plants in different stages of development to determine the real condition of the lamellæ.

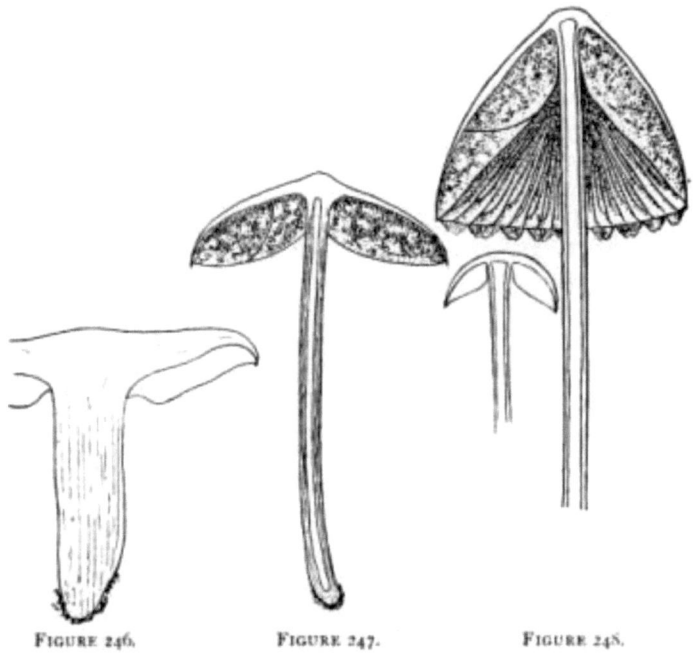

FIGURE 246.   FIGURE 247.   FIGURE 248.

- Figure 246. — Tricholoma, gills sinuate, stipe solid.
- Figure 247. — Panæolus papilionaceus, gills adnate.
- Figure 248. — Left-hand small plant, Hygrophorus, gills sinuate, adnexed. Right-hand plant Panæolus retirugis, gills ascending adnexed, veil appendiculate.

In certain genera the gills have special characteristics which may be noted here. Usually the edge of the lamellæ is *acute* or sharp like the blade of a knife, but in *Cantharellus* and *Trogia* the edges are very blunt or obtuse. In extreme forms the lamellæ are reduced to [Pg 303] mere veins or ridges. Again, the edge is generally *entire*, i. e., not noticeably toothed, but in *Lentinus* it is often toothed or cut in various ways. In some other plants the edges are *serrulate, crenulate,* etc. In *Schizophyllum alneum,* a small whitish plant very common on dead sticks, the gills are split lengthwise along the edge with the halves revolute, i. e., rolled back. In *Coprinus* the gills and often a large part of the pileus melt at maturity into a dark, inky fluid.

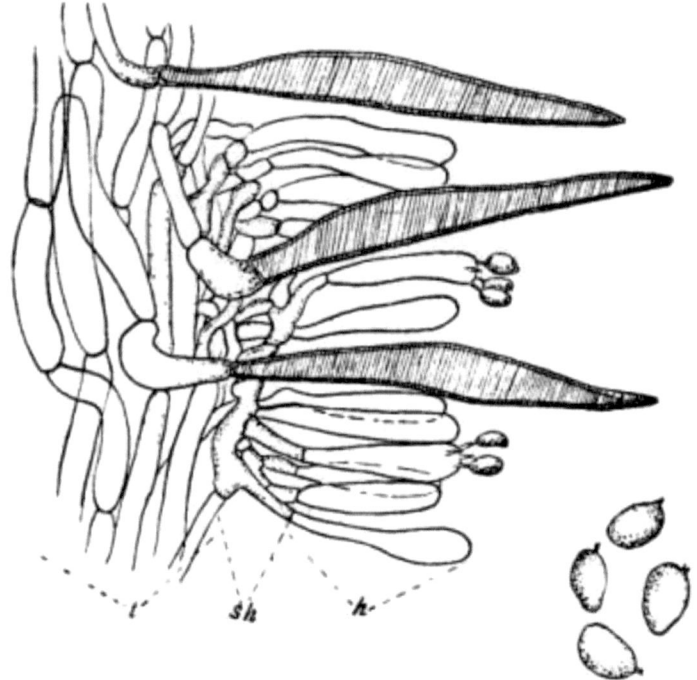

Figure 249.—Section of portion of gill of Marasmius cohærens. *t*, trama of gill; *sh*, sub-hymenium; *h*, hymenium layer. The long, dark cells are brown cystidia, termed spicules by some to distinguish them from the colorless cystidia. The long cells bearing the oval spores are the basidia.

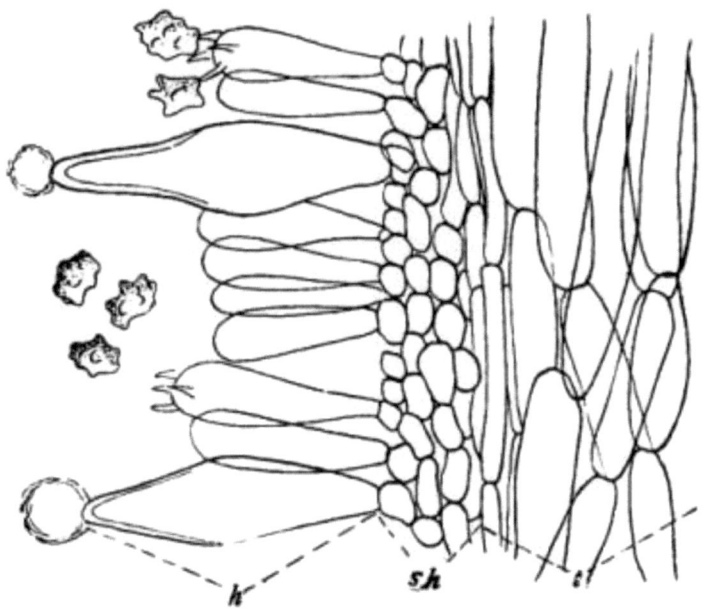

Figure 250.—Inocybe repanda (Bull.) Bres. (= Entoloma repandum Bull.). *t*, trama of pileus; *sh*, sub-hymenium; *h*, the hymenial layer; the long cells with a drop of moisture at the ends are cystidia (sing. cystidium).

The hymenium.—The term *hymenium* is applied to the spore-bearing tissue of many fungi. In the *Agaricaceæ* the hymenium covers the entire surface of the gills and usually the portion of the pileus between the gills. It originates in the following manner: the threads forming the trama of the gills grow out from the lower side of the pileus and perpendicular to its under surface. As growth advances many branches of the threads turn outward toward either surface of the gill and finally terminate in club-shaped cells. These cells, therefore, lie side by side, perpendicular to the surface, [Pg 304] forming a pavement, as it were, over the entire surface of the gills. Some of them put out four little prongs, on each of which a spore is borne, while others simply remain as sterile cells (Figs. 249, 250). The spore-bearing cells are *basidia*; the others are called *paraphyses*. They resemble each other very much, except that the basidia bear four *sterigmata* and a spore on each. In a few species the number of sterigmata is reduced to two and in some low forms the number is variable. The layer just beneath the basidia is usually more or less modified, being often composed of small cells different from the rest of the trama. This is called the *sub-hymenial* layer or *sub-hymenium* (Fig. 250).

Other cells called *cystidia* occur in the hymenia of various species distributed through nearly all the genera of the agarics. Cystidia are large, usually inflated, cells which project above the rest of the hymenium (Fig. 250). They originate either like the basidia, from the sub-hymenial cells (Fig. 250), or from special hyphæ deeper down in the trama of the gill (Fig. 249). They are scattered over the entire surface of the hymenium, but become more numerous on the edge of the lamellæ. Their number is much smaller than that of the basidia, but in some species where they are colored they may greatly change the appearance of the gills. Cystidia often secrete moisture which collects in drops at their tips, a phenomenon common to all free fungous cells.

The stem. — The stem is usually fixed to the center of the pileus, but it may be *eccentric*, i. e., fixed to one side of the center, or entirely lateral. When the stem is wanting the pileus is *sessile*. With regard to its interior the stem is *solid*, when it is evenly fleshy throughout (Fig. 246), or *hollow* when the interior is occupied by a cavity (Fig. 248). If the cavity is narrow and tubular the stem is *fistulose* (Fig. 245); and if the center is filled with a pithy substance it is *stuffed* (Fig. 243). These terms apply only to the natural condition of the stem, and not the condition brought about by larvæ, which eat out the interior of the stem, causing it to be hollow or fistulose.

The terms applicable to the consistency of the stem are difficult to define. In general, stems may be either *fleshy* or *cartilaginous*. The meaning of these terms can best be learned by careful study of specimens of each, but a few general characters can be given here. Fleshy, fibrous stems occur in the genera *Clitocybe* and *Tricholoma*, among the white-spored forms. Their consistency is like that of the pileus, namely, made up of fleshy, fibrous tissue. They are usually stout, compared with the size of the plant, and when bent or broken [Pg 305] they seem to be more or less spongy or tough, fibrous, so that they do not snap readily. Cartilaginous stems have a consistency resembling that of cartilage. Their texture is always different from that of the pileus, which is fleshy or membranous. In general such stems are rather slender, in many genera rather thin, but firm. When bent sufficiently they either snap suddenly, or break like a green straw, without separating. In regard to their external appearance some resemble fibrous stems, while others are smooth and polished as in *Mycena* and *Omphalia*.

The veil. — In the young stages of development the margin of the pileus lies in close contact with the stipe, the line of separation being indicated by a kind of furrow which runs around the young button mushroom. In many genera, as *Collybia*, *Mycena*, *Omphalia*, etc., the pileus simply expands without having its margin ever united to the stipe by any special structure, but in other forms, which include by far the greater number of genera of the *Agaricaceæ* and some *Boleti*, the interval between the stem and pileus is bridged over by threads growing from the margin of the pileus and from the outer layers of the stem. These threads interlace to form a delicate membrane, known as the *veil*, which closes the gap between the stem and pileus and covers over the young hymenium.

The veil remains firm for a time, but it is finally torn by the expanding pileus, and its remnants persist on the cap and stem in the form of various appendages, whose character depends on the character of the veil. In *Cortinarius* the veil is made up of delicate threads extending radially from the stem to the margin of the cap without forming a true membrane. From its resemblance to a spider's web such a veil is said to be *arachnoid*. At maturity mere traces of it can be found on the stem. In many genera the veil consists of a delicate membrane which tears away from the stem and hangs in flakes to the margin of the pileus. In these cases the veil is *appendiculate* (Fig. 248). Frequently it is so delicate that no trace of it remains on the mature plant. Where the veil is well developed it usually remains on the stem as a *ring* or *annulus* which becomes free and movable in species of *Lepiota* (Fig. 242) and *Coprinus*, or forms a hanging annular curtain in *Amanita*, or a thick, felty ring in *Agaricus*, etc. In some plants (species of *Lepiota*) the annulus is continuous with the outer cortex of the stem, which then appears as if it were partially enclosed in a sheath, with the annulus forming a fringe on the upper end of the sheath, from which the apex of the stem projects.

No reference is here made to the *volva*, which encloses the entire [Pg 306] plant, and which is described in connection with the genera in which it occurs.

The few typical characters described here will help the student to become familiar with terms applied to them. In nature, however, typical cases rarely exist, and it is often necessary to draw distinction between differences so slight that it is almost impossible to describe them. Only by patient study and a thorough acquaintance with the characters of each genus can one hope to become familiar with the many mushrooms growing in our woods and fields.

CHAPTER XXIV.

ANALYTICAL KEYS.

By the Author.

CLASS FUNGI.

SUB-CLASS BASIDIOMYCETES. [F]

Plants of large or medium size; fleshy, membranaceous, leathery, woody or gelatinous; growing on the ground, on wood or decaying organic matter; usually saprophytic, more rarely parasitic. Fruiting surface, or hymenium, formed of numerous crowded perpendicular basidia, the apex of the latter bearing two to six (usually four) basidiospores, or the basidiospores borne laterally; in many cases cystidia intermingled with the basidia. Hymenium either free at the beginning, or enclosed either permanently or temporarily in a more or less perfect peridium or veil. Basidiospores continuous or rarely septate, globose, obovoid, ellipsoidal to oblong, smooth or roughened, hyaline or colored, borne singly at the apex of sterigmata.

Order *Gasteromycetes*. Plants membranaceous, leathery or fleshy, furnished with a peridium and gleba, the latter being sometimes supported on a receptacle. Hymenium on the surface of the gleba which is enclosed within the peridium up to the maturity of the spores or longer; spores continuous, sphæroid or ellipsoid, hyaline or colored. Puff-balls, etc.

Order *Hymenomycetes*. Hymenium, at the beginning, borne on the free outer surface of the compound sporophore, or if at first enclosed by a pseudo-peridium or veil it soon becomes exposed before the maturity of the spores; mushrooms, etc. [Pg 307]

HYMENOMYCETES.

Analytical Key of the Families.

	Plants not gelatinous; basidia continuous.	1
	Plants gelatinous or sub-gelatinous, basidia forked, or divided longitudinally or transversely.	4
1 —	Hymenium uneven, i. e., in the form of radiating plates, or folds; or a honey-combed surface, or reticulate, warty, spiny, etc.	2
	Hymenium smooth (not as in B, though it may be convolute and irregular, or ribbed, or veined).	3
2 —	Hymenium usually on the under side, in the form of radiating plates, or strong folds. The genus Phlebia in the Hydnaceae has the hymenium on smooth, somewhat radiating veins which are interrupted and irregular. One exotic genus has the hymenium on numerous irregular obtuse lobes (Rhacophyllus).	**Agaricaceæ.** 17
	Hymenium usually below (or on the outer surface when the plant is spread over the substratum), honey-combed, porous, tubulose, or reticulate; in one genus with short, concentric plates.	**Polyporaceæ.** 171

Hymenium usually below (or on the outer surface when the plant is spread over the substratum), warted, tuberculate, or with stout, spinous processes; or with interrupted vein-like folds in resupinate forms.	**Hydnaceæ.**	195
3 — Plants somewhat corky or membranaceous, more or less expanded; hymenium on the under surface (upper surface sterile), or on the outer or exposed surface when the plant is spread over the substratum (margin may then sometimes be free, but upper surface, i. e., that toward the substratum, sterile). (Minute slender spines are sometimes intermingled with the elements of the hymenium, and should not be mistaken for the stouter spinous processes of the Hydnaceæ).	**Thelephoraceæ.**	208
Plants more or less fleshy, upright (never spread over the surface of the substratum), simple or branched. Hymenium covering both sides and the upper surface.	**Clavariaceæ.**	200
4 — Basidia forked or longitudinally divided; or if continuous then globose, or bearing numerous spores; or if the plant is leathery, membranous, or floccose, then basidia as described. Hymenium covering the entire free surface or confined to one portion; smooth, gyrose, folded or lobed; or hymenium lamellate, porous, reticulate or toothed forms which are gelatinous and provided with continuous basidia may be sought here.	**Tremellineæ.**	204

[Pg 308]

FAMILY AGARICACEAE.

Pileus more or less expanded, convex, bell-shaped; stipe central or nearly so; or the point of attachment lateral, when the stipe may be short or the pileus sessile and shelving. Fruiting surface usually on the under side and exposed toward the earth, lamellate, or prominently folded or veined. Lamellæ or gills radiating from the point of attachment of the pileus with the stipe or with the substratum in the sessile forms; lamellæ simple or branched, rarely anastomosing behind, clothed externally on both surfaces with the basidia, each of which bears four spores (rarely two), cystidia often present.

Key to the North American genera.

THE WHITE-SPORED AGARICS.

(Sometimes there is a faint tinge of pink or lilac when the spores are in bulk, but the color is not seen under the microscope.)

	Plants soft, fleshy or nearly so, usually soon decaying; dried plants do not revive well when moistened.	1
	Plants tough, either fleshy or gelatinous, membranaceous, corky or woody, persistent, reviving when moistened.	13
1—	**Gills** acute on the edge.	2
	Edge of the gills obtuse, or gills fold-like, or vein-like, but prominent.	12
2—	**Trama** of the pileus of interwoven threads, not vesiculose.	3
	Trama of the pileus vesiculose, plants rigid but quite fragile.	11
3—	**Gills** thin, not much broadened toward the pileus.	4
	Gills broadened toward the pileus, of waxy consistency.	**Hygrophorus.** 110

4 —	**Stipe** central or sub-central. (Some species of Pleurotus are sub-central, but the gills are usually not decurrent.)	5
	Stipe on one side of the pileus, or none, rarely with the stipe sub-central. (Some species of Clitocybe are sub-central.)	**Pleurotus.** 102
5 —	**Stipe** fleshy, pileus easily separating from the stipe, gills usually free.	6
	Stipe fleshy or fibrous and elastic, pileus confluent with the stipe and of the same texture.	7
	Stipe cartilaginous, pileus confluent with the stipe, but of a different texture.	9
6 —	**Volva** and annulus present on the stipe.	**Amanita.** 52
	Volva present, annulus wanting.	**Amanitopsis.** 74
	Volva wanting, annulus present.	**Lepiota.** 77
7 —	**Annulus** and volva wanting.	8
	Annulus usually present (sometimes vague), volva wanting, gills attached to the stipe.	**Armillaria.** 83
8 —	**Gills** sinuate.	**Tricholoma.** 87
	Gills decurrent, not sinuate.	**Clitocybe.** 89
9 —	**Gills** decurrent, pileus umbilicate.	**Omphalia.** 100
	Gills not decurrent.	10
[Pg 309] 10 —	**Margin** of pileus at first involute, pileus flat or nearly so, somewhat fleshy (some plants rather tough and	**Collybia.** 92

460

tending toward the consistency of Marasmius).

Margin of the pileus at first straight, pileus slightly bell-shaped, thin.	Mycena.	93
Gills usually free, pileus deeply plicate so that the gills are split where they are attached to the pileus, pileus membranaceous, very tender but not diffluent.	Hiatula.	
11 — **Plants** where bruised exuding a milky or colored juice.	Lactarius.	114
Plants not exuding a juice where bruised.	Russula.	125
12 — **Gills** decurrent, dichotomous, edge blunt.	Cantharellus.	128
Gills not decurrent, plants parasitic on other mushrooms.	Nyctalis.	
13 — **Edge** of gills not split into two laminæ.	14	
Edge of gills split into two laminæ and revolute.	Schizophyllum.	136
14 — **Plants** leathery, either fleshy, membraneous, or gelatinous.	15	
Plants corky or woody (placed by some in Polyporaceæ).	Lenzites.	
15 — **Stipe** separate from the pileus (hymenophore), easily separating.	16	
Stipe continuous with hymenophore.	17	
16 — **Plants** tough and fleshy, membranaceous or leathery.	Marasmius.	130
Plants gelatinous and leathery.	Heliomyces.	

17 — **Edge** of the gills acute.	18	
Edge of the gills obtuse.	19	
18 — **Edge** of gills usually serrate.	**Lentinus.**	134
Edge of gills entire.	**Panus.**	134
19 — **Gills** dichotomous.	**Xerotus.**	
Gills fold-like, irregular.	**Trogia.**	137

There are only a few rare species of Hiatula, Nyctalis, Heliomyces and Xerotus in the United States. None are here described.

THE OCHRE-SPORED AGARICS.

(The spores are yellowish brown or rusty brown.)

Gills not separating readily from	1	
Gills sometimes separating readily from the pileus, forked or anastomosing at the base, or connected with vein-like reticulations.	**Paxillus.**	165
1 — **Universal veil** not arachnoid (i. e., not cobwebby).	2	
Universal veil arachnoid, distinct from the cuticle of the pileus, gills powdery from the spores.	**Cortinarius.**	161
2 — **Stipe** central.	3	
Stipe eccentric or none.	**Crepidotus.**	159
3 — **Volva** or annulus present on stipe.	4	
Volva and annulus wanting.	5	
4 — **Stipe** with an annulus.	**Pholiota.**	150
Stipe with a volva. **Locellina** (not reported in U. S.).		
5 — **Gills** free from the stem.	**Pluteolus.**	

	Gills attached.	6	
6 —	**Gills** not dissolving nor becoming powdery.	7	
[Pg 310]	**Gills** dissolving into a gelatinous or powdery condition, not diffluent as in Coprinus.	**Bolbitius.**	163
7 —	**Stipe** fleshy.	8	
	Stipe cartilaginous or sub-cartilaginous.	10	
8 —	**Gills** somewhat sinuate.	9	
	Gills adnate or decurrent.	**Flammula.**	156
9 —	**Cuticle** of the pileus silky or bearing fibrils.	**Inocybe.**	158
	Cuticle of pileus smooth, viscid.	**Hebeloma.**	157
10 —	**Gills** decurrent.	**Tubaria.**	159
	Gills not decurrent.	11	
11 —	**Margin** of pileus inflexed.	**Naucoria.**	153
	Margin of pileus straight, from the first.	**Galera.**	155

No species of Pluteolus are here described.

THE ROSY-SPORED AGARICS.

(The spores are rose color, pink, flesh or salmon color.)

	Stipe central.	1	
	Stipe eccentric or none and pileus lateral.	**Claudopus.**	149
1 —	**Pileus** easily separating from the stipe, gills free.	2	
	Pileus confluent with the stipe and of the same texture, gills attached, in some becoming almost free.	3	

2 —	**Volva** present and distinct, annulus wanting.	**Volvaria.**	140
	Volva and annulus wanting.	**Pluteus.**	138
3 —	**Stipe** fleshy to fibrous, margin of pileus at first incurved.	4	
	Stipe cartilaginous.	5	
4 —	**Gills** sinuate.	**Entoloma.**	143
	Gills decurrent.	**Clitopilus.**	142
5 —	**Gills** not decurrent (or if so only by a minute tooth), easily separating from the stipe.	6	
	Gills decurrent, pileus umbilicate.	**Eccilia.**	148
6 —	**Pileus** slightly convex, margin at first incurved.	**Leptonia.**	147
	Pileus bell-shaped, margin at first straight and pressed close against the stipe.	**Nolanea.**	

No species of Nolanea are described here.

THE BROWN-SPORED AGARICS.

(The spores are dark brown or purplish brown.)

	Pileus easily separating from the stem; gills usually free.	1	
	Pileus continuous with the stem; gills attached.	2	
1 —	**Volva** wanting, annulus present. (Psalliota Fr.)	**Agaricus.**	18
	Volva present, annulus wanting.	**Chitonia.**	
	Volva and annulus wanting.	**Pilosace.**	
2 —	**Veil** present.	3	
	Veil wanting or obsolete.	4	

3—	**Annulus** present, gills attached.	**Stropharia.**	31
	Annulus wanting, veil remaining attached to margin of pileus.	**Hypholoma.**	26
4—	**Stipe** <u>tenacious</u>, margin of pileus first incurved.	5	
	Stipe fragile, margin of pileus at first straight.	**Psathyra.**	
[Pg 311] 5—	**Gills** sub-triangularly decurrent.	**Deconica.**	
	Gills not decurrent.	**Psilocybe.**	

But few species of Psathyra, Deconica, Chitonia and Pilosace are noted from the United States. None are here described.

THE BLACK-SPORED AGARICS.

	Pileus present to which the gills are attached.	1	
	Pileus wanting, gills attached to a disk at apex of stem from which they radiate.	**Montagnites.** [G]	
1—	**Gills** more or less deliquescing, or pileus thin, membranous and splitting between the laminæ of the gills and becoming more or less plicate.	**Coprinus.**	32
	Gills not deliquescing, etc.	2	
2—	**Spores** globose, ovoid.	3	
	Spores elongate, fusiform (in some species brown), plants with a slimy envelope.	**Gomphidius.**	49
3—	**Pileus** somewhat fleshy, not striate, projecting beyond the gills at the margin; gills variegated in color from groups of dark spores on the surface.	4	
	Pileus somewhat fleshy, margin striate,	**Psathyrella.**	48

gills not variegated.

4 —	**Annulus** wanting, but veil often present.	**Panæolus.**	45
	Annulus wanting, veil appendiculate on margin of cap.	**Chalymotta.**	48
	Annulus present.	**Anellaria.**	

GLOSSARY OF THE MORE TECHNICAL TERMS USED IN THIS WORK.

Abbreviations:

- cm. = centimeter (about 2-1/2 cm. make one inch).
- mm. = millimeter (about 25 mm. make 1 inch).
- μ = one micron (1000 μ = 1 mm.).

Adnate, said of the gills when they are attached squarely, or broadly, to the stem.

Adnexed, said of gills when they are attached only slightly or only by the upper angle of the stem.

Anastomose, running together in a net-like manner.

Annulus, the ring or collar around the stem formed from the inner or partial veil.

Appendiculate, said of the veil when it clings in fragments to the margin of the pileus.

Arachnoid, said of the veil when it is cobwebby, that is, formed of loose threads.

Ascus, the club-shaped body which bears the spores inside (characteristic of the Ascomycetes).

Basidium (pl. basidia) the club-shaped body which bears the spores in the Basidiomycetes. These stand parallel, and together make up the entire or large part of the hymenium or fruiting surface which covers the gills, etc. Paraphyses (sterile cells) and sometimes cystidia (longer sterile cells) or spines are intermingled with the basidia.

Bulbous, said of the enlarged lower end of the stem in some mushrooms.

Circumscissile, splitting transversely across the middle, used to indicate one of the ways in which the volva ruptures. [Pg 312]

Cortina, a cobwebby veil.

Cuticle, the skin-like layer on the outside of the pileus.

Decurrent, said of the gills when they extend downward on the stem.

Diffluent, said of the gills when they dissolve into a fluid.

Dimidiate, halved, said of a sessile pileus semi-circular in form and attached by the plane edge directly to the wood.

Echinulate, term applied to minute spinous processes, on the spores for example.

Eccentric, said of a stem when it is attached to some other point than the center of the pileus.

Fimbriate, in the form of a delicate fringe.

Fistulose, becoming hollow.

Floccose, term applied to indicate delicate and soft threads, cottony extensions from the surface of any part of the mushroom.

Flocculose, minutely floccose.

Fugacious, disappearing.

Fuliginous (or fuligineous), dark brown, sooty or smoky.

Fulvous, tawny, reddish yellow.

Fusiform, spindle-shaped.

Fusoid, like a spindle.

Furfuraceous, with numerous minute scales.

Gleba, the chambered tissue forming the hymenium (fruiting surface) in the puff-balls and their allies.

Hygrophanous, appearing to be water soaked.

Hymenium, the fruiting surface of the mushrooms and other fungi.

Hymenomycetes, the subdivision of the Basidiomycetes in which the fruiting surface is exposed before the spores are ripe.

Hymenophore, the portion of the fruit body which bears the hymenium.

Hypha (pl. hyphæ), a single mycelium thread.

Imbricate, overlapping like the shingles on a roof.

Involute, folded or rolled inward.

Lamella (pl. lamellæ), the gills of the mushroom.

Mycelium, the vegetative or growing portion of the mushrooms, and other fungi, made up of several or many threads.

Ocreate, applied to the volva where it fits the lower part of the stem, as a stocking does the leg.

Pectinate, like the teeth of a comb.

Peridium, the wall of the puff-balls, etc.

Pileus (pl. pilei), the cap of the mushroom.

Plicate, plaited, or folded like a fan.

Punctate, with minute points.

Pulverulent, with a minute powdery substance.

Repand, wavy.

Resupinate, spread over the matrix, the fruiting surface external and the pileus next the wood.

Revolute, rolled backward.

Rugose, wrinkled.

Rugulose, with minute wrinkles.

Saprophytic, growing on dead organic matter.

Sessile, where the pileus is attached directly to the matrix without any stem.

Sinuate, said of the gills when they are notched at their junction with the stem. [Pg 313]

Stipe, the stem.

Sulcate, furrowed.

Squamulose, with minute scales.

Squarrose, with prominent reflexed scales.

Tomentose, with a dense, matted, hairy or woolly surface.

Trama, the interior portion of the gills or pileus.

Umbo, with a prominent boss or elevation, in the center of the pileus.

Umbilicate, with a minute abrupt depression in the center of the cap.

Veil, a layer of threads extending from the margin of the cap to the stem (partial veil or marginal veil). A universal veil envelops the entire plant.

Veins, elevated lines or folds running over the surface of the lamellæ in some species, and often connected so as to form reticulations.

Ventricose, enlarged or broadened at the middle, bellied.

Vesiculose, full of small rounded vesicles, as the trama of the pileus of a Russula.

Volva, a wrapper or envelope, which in the young stage completely surrounds the plant, same as universal veil. At maturity of the plant it may be left in the form of a cup at the base of the stem, or broken up into fragments and distributed over the cap and base of the stem.

FOOTNOTES:

[F] The sub-class Ascomycetes includes the morels, helvellas, cup fungi, etc., and many microscopic forms, in which the spores are borne inside a club-shaped body, the ascus. Only a few of the genera are described in this book, and the technical diagnosis will be omitted. See page 216.

[G] One American species in Texas.

[Pg 314]

[Pg 315] INDEX OF GENERA, AND ILLUSTRATIONS.

Note.—In this index the generic and specific names have been divided into syllables, and the place of the primary accent has been indicated, with the single object of securing a uniform pronunciation in accordance with the established rules of English orthoepy.

470

- Volvaria (vol-va'ri-a), 140.
    - bombycina (bom-byc'i-na), 140, 141, fig. 137.
    - speciosa (spe-ci-o'sa), 141, 142.

- White-spored agarics, 52.

INDEX TO SPECIES.

- esculenta (Gyromitra), 220.
- esculenta (Morchella), 217.

- farinosa (Amanitopsis), 76.
- felleus (Boletus), 173.
- fimicola (Panæolus), 48.
- firma (Fistulina), 186.
- flava (Galera), 155.
- flavidus (Boletus), 178.
- flavidus (Paxillus), 168.
- floccocephala (Amanita), 62.
- floccopus (Strobilomyces), 185.
- floccosa (Sarcoscypha), 221.
- fœnisecii (Psilocybe), 48.
- fomentarius (Polyporus), 194.
- formosa (Clavaria), 201.
- fragile (Hydnum), 200.
- fragilis (Russula), 127.
- frondosa (Tremella), 205.
- frondosus (Polyporus), 188.
- frostiana (Amanita), 54.
- fuciformis (Tremella), 206.
- fuligineus (Hygrophorus), 113.
- fuliginosus (Lactarius), 118.
- fulvotomentosus (Crepidotus), 161.
- furcata (Russula), 127.
- fusiger (Spinellus), 95.

- galericulata (Mycena), 94.
- gemmatum (Lycoperdon), 210.
- gerardii (Lactarius), 119.
- giganteum (Lycoperdon), 210.
- glandulosa (Exidia), 206.
- glutinosus (Gomphidius), 51.
- granulatus (Boletus), 178.
- graveolens (Hydnum), 200.